*f*P

FOR THE LOVE
OF PHYSICS

From the End of the Rainbow to the Edge of Time—
A Journey Through the Wonders of Physics

WALTER LEWIN

with

WARREN GOLDSTEIN

Free Press

New York London Toronto Sydney

*f*P
Free Press
A Division of Simon & Schuster, Inc.
1230 Avenue of the Americas
New York, NY 10020

First Free Press hardcover edition May 2011

FREE PRESS and colophon are trademarks of Simon & Schuster, Inc.

For information about special discounts for bulk purchases, please contact Simon & Schuster Special Sales at 1-866-506-1949 or business@simonandschuster.com.

The Simon & Schuster Speakers Bureau can bring authors to your live event. For more information or to book an event contact the Simon & Schuster Speakers Bureau at 1-866-248-3049 or visit our website at www.simonspeakers.com.

Book design by Ellen R. Sasahara

Manufactured in the United States of America

3   5   7   9   10   8   6   4   2

Library of Congress Cataloging-in-Publication Data
Lewin, Walter H. G.
For the love of physics : from the end of the rainbow to the edge of time—a journey through the wonders of physics / by Walter Lewin with Warren Goldstein.
p.   cm.
1. Lewin, Walter H. G.   2. Physicists—Massachusetts—Biography.   3. College teachers—Massachusetts—Biography. 4. Physics—Study and teaching—Netherlands. 5. Physics—Study and teaching—Massachusetts.
I. Goldstein, Warren Jay.   II. Title.
QC16.L485A3 2011
530.092—dc22
[B]   2010047737

ISBN 978-1-4391-0827-7
ISBN 978-1-4391-2354-6 (ebook)

For all who inspired my love for physics and art
—Walter Lewin

For my grandson Caleb Benjamin Luria
—Warren Goldstein

CONTENTS

INTRODUCTION

S ix feet two and lean, wearing what looks like a blue work shirt, sleeves rolled to the elbows, khaki cargo pants, sandals and white socks, the professor strides back and forth at the front of his lecture hall, declaiming, gesturing, occasionally stopping for emphasis between a long series of blackboards and a thigh-high lab table. Four hundred chairs slope upward in front of him, occupied by students who shift in their seats but keep their eyes glued to their professor, who gives the impression that he is barely containing some powerful energy coursing through his body. With his high forehead, shock of unruly grey hair, glasses, and the hint of some unidentifiable European accent, he gives off a hint of Christopher Lloyd's Doc Brown in the movie *Back to the Future*—the intense, otherworldly, slightly mad scientist-inventor.

But this is not Doc Brown's garage—it's the Massachusetts Institute of Technology, the preeminent science and engineering university in the United States, perhaps even the world, and lecturing at the blackboard is Professor Walter H. G. Lewin. He halts his stride and turns to the class. "Now. All important in making measurements, which is *always* ignored in *every* college physics book"—he throws his arms wide, fingers spread—"is the uncertainty in your measurements." He pauses, takes a step, giving them time to consider, and stops again: "Any measurement that you make without knowledge of the uncertainty is *meaningless.*" And the hands fly apart, chopping the air for emphasis. Another pause.

"I will repeat this. I want you to hear it tonight at three o'clock in the morning when you wake up." He is holding both index fingers to his temples, twisting them, pretending to bore into his brain. "Any measurement that you make without knowledge of its uncertainty is completely *meaningless.*" The students stare at him, utterly rapt.

We're just eleven minutes into the first class of Physics 8.01, the most famous introductory college physics course in the world.

The *New York Times* ran a front-page piece on Walter Lewin as an MIT "webstar" in December 2007, featuring his physics lectures available on the MIT OpenCourseWare site, as well as on YouTube, iTunes U, and Academic Earth. Lewin's were among the first lectures that MIT posted on the Internet, and it paid off for MIT. They have been exceptionally popular. The ninety-four lectures—in three full courses, plus seven stand-alones—garner about three thousand viewers per day, a million hits a year. Those include quite a few visits from none other than Bill Gates, who's watched all of courses 8.01, Classical Mechanics, and 8.02, Electricity and Magnetism, according to letters (snail mail!) he's sent Walter, reporting that he was looking forward to moving on to 8.03, Vibrations and Waves.

"You have changed my life," runs a common subject line in the emails Lewin receives every day from people of all ages and from all over the world. Steve, a florist from San Diego, wrote, "I walk with a new spring in my step and I look at life through physics-colored eyes." Mohamed, an engineering prep school student in Tunisia wrote, "Unfortunately, here in my country my professors don't see any beauty in physics as you do see, and I've suffered a lot from this. They just want us to learn how to solve 'typical' exercises to succeed in the exam, they don't look beyond that tiny horizon." Seyed, an Iranian who had already earned a couple of American master's degrees, writes, "I never really enjoy of life until I have watched you teach physics. Professor Lewin you have changed my life Indeed. The way you teach it is worth 10 times the tuition, and make

SOME not all other teachers bunch of criminals. It is CAPITAL CRIME to teach bad." Or Siddharth from India: "I could feel Physics beyond those equations. Your students will always remember you as I will always remember you—as a very-very fine teacher who made life and learning more interesting than I thought was possible."

Mohamed enthusiastically quotes Lewin's final lecture in Physics 8.01 with approval: "Perhaps you will always remember from my lectures that physics can be very exciting and beautiful and it's everywhere around us, all the time, if only you have learned to see it and appreciate its beauty." Marjory, another fan, wrote, "I watch you as often as I can; sometimes five times per week. I am fascinated by your personality, your sense of humor, and above all by your ability to simplify matters. I hated physics in high school, but you made me love it."

Lewin receives dozens of such emails every week, and he answers each one.

Walter Lewin creates magic when he introduces the wonders of physics. What's his secret? "I introduce people to their own world," he says, "the world they live in and are familiar with, but don't approach like a physicist—yet. If I talk about waves on water, I ask them to do certain experiments in their bathtubs; they can relate to that. They can relate to rainbows. That's one of the things I love about physics: you get to explain anything. And that can be a wonderful experience—for them and for me. I make them love physics! Sometimes, when my students get really engaged, the classes almost feel like happenings."

He might be perched at the top of a sixteen-foot ladder sucking cranberry juice out of a beaker on the floor with a long snaking straw made out of lab tubing. Or he could be courting serious injury by putting his head in the path of a small but quite powerful wrecking ball that swings to within millimeters of his chin. He might be firing a rifle into two paint cans filled with water, or charging himself with 300,000 volts of electricity with a large contraption called a Van de Graaff generator—like something out of a mad scientist's laboratory in a science fiction movie—so that his already wild hair stands straight out from his skull. He uses his body as a piece of experimental equipment. As he says often, "Science requires sacrifices, after all." In one demonstration—captured in the photo on the jacket of this book—he sits on an extremely uncomfortable metal ball at the end of a rope suspended from the lecture hall's ceiling (what he calls the mother of all pendulums) and swings back and forth while his students chant the number of swings, all to prove that the number of swings a pendulum makes in any given time is independent of the weight at its end.

His son, Emanuel (Chuck) Lewin, has attended some of these lectures and recounts, "I saw him once inhale helium to change his voice. To get the effect right—the devil is in the details—he typically gets pretty close to the point of fainting." An accomplished artist of the blackboard, Lewin draws geometrical figures, vectors, graphs, astronomical phenomena, and animals with abandon. His method of drawing dotted lines so entranced several students that they produced a funny YouTube video titled "Some of Walter Lewin's Best Lines," consisting simply of lecture excerpts showing Lewin drawing his famous dotted lines on different blackboards during his 8.01 lectures. (You can watch it here: www.youtube.com/watch?v=raurl4s0pjU.)

A commanding, charismatic presence, Lewin is a genuine eccentric: quirky and physics obsessed. He carries two devices called polarizers in his wallet at all times, so that at a moment's notice he can see if any source of light, such as the blue sky, a rainbow, or reflections off windows, is polarized, and whoever he might be with can see it too.

What about those blue work shirts he wears to class? Not work shirts at all, it turns out. Lewin orders them, custom made to his specifications, of high-grade cotton, a dozen at a time every few years, from a tailor in Hong Kong. The oversize pocket on the left side Lewin designed to accommodate his calendar. No pocket protectors here—this physicist-performer-teacher is a man of meticulous fashion—which makes a person wonder why he appears to be wearing the oddest brooch ever worn by a university professor: a plastic fried egg. "Better," he says, "to have egg on my shirt than on my face."

What is that oversize pink Lucite ring doing on his left hand? And what is that silvery thing pinching his shirt right at belly-button level, which he keeps sneaking looks at?

Every morning as Lewin dresses, he has the choice of forty rings and thirty-five brooches, as well as dozens of bracelets and necklaces. His taste runs from the eclectic (Kenyan beaded bracelets, a necklace of large amber pieces, plastic fruit brooches) to the antique (a heavy silver Turkmen cuff bracelet) to designer and artist-created jewelry, to the simply and hilariously outrageous (a necklace of felt licorice candies). "The students started noticing," he says, "so I began wearing a different piece every lecture. And especially when I give talks to kids. They love it."

And that thing clipped to his shirt that looks like an oversize tie clip? It's a specially designed watch (the gift of an artist friend) with the face upside down, so Lewin can look down at his shirt and keep track of time.

It sometimes seems to others that Lewin is distracted, perhaps a classic absentminded professor. But in reality, he is usually deeply engaged in thinking about some aspect of physics. As his wife Susan Kaufman recently recalled, "When we go to New York I always drive. But recently I took this map out, I'm not sure why, but when I did I noticed there were equations all over the margins of the states. Those margins were done when he was last lecturing, and he was bored when we were driving. Physics was always on his mind. His students and school were with him twenty-four hours a day."

Perhaps most striking of all about Lewin's personality, according to his longtime friend the architectural historian Nancy Stieber, is "the laser-sharp intensity of his interest. He seems always to be maximally engaged in whatever he chooses to be involved in, and eliminates 90 percent of the world. With that laserlike focus, he eliminates what's inessential to him, getting to a form of engagement that is so intense, it produces a remarkable *joie de vivre*."

Lewin is a perfectionist; he has an almost fanatical obsession with detail. He is not only the world's premier physics teacher; he was also a pioneer in the field of X-ray astronomy, and he spent two decades building, testing, and observing subatomic and astronomical phenomena with ultrasensitive equipment designed to measure X-rays to a remarkable degree of accuracy. Launching enormous and extremely delicate balloons that skimmed the upper limit of Earth's atmosphere, he began to uncover an exotic menagerie of astronomical phenomena, such as X-ray bursters. The discoveries he and his colleagues in the field made helped to demystify the nature of the death of stars in massive supernova explosions and to verify that black holes really do exist.

He learned to test, and test, and test again—which not only accounts for his success as an observational astrophysicist, but also for the remarkable clarity he brings to revealing the majesty of Newton's laws, why the strings of a violin produce such beautifully resonant notes, and why you lose and gain weight, be it only very briefly, when you ride in an elevator.

For his lectures, he always practiced at least three times in an empty classroom, with the last rehearsal being at five a.m. on lecture day. "What makes his lectures work," says astrophysicist David Pooley, a former student who worked with him in the classroom, "is the time he puts into them."

When MIT's Physics Department nominated Lewin for a prestigious teaching award in 2002, a number of his colleagues zeroed in on these exact qualities. One of the most evocative descriptions of the experience of learning physics from Lewin is from Steven Leeb, now a professor of electrical engineering and computer science at MIT's Laboratory for Electromagnetic and Electronic Systems, who took his Electricity and Magnetism course in 1984. "He exploded onto the stage," Leeb recalls, "seized us by the brains, and took off on a roller-coaster ride of electromagnetics that I can still feel on the back of my neck. He is a genius in the classroom with an unmatched resourcefulness for finding ways to make concepts plain."

Robert Hulsizer, one of Lewin's Physics Department colleagues, tried to excerpt some of Lewin's in-class demonstrations on video to make a kind of highlight film for other universities. He found the task impossible. "The demonstrations were so well woven into the development of the ideas, including a buildup and denouement, that there was no clear time when the demonstration started and when it finished. To my mind, Walter had a richness of presentation that could not be sliced into bites."

The thrill of Walter Lewin's approach to introducing the wonders of physics is the great joy he conveys about all the wonders of our world. His son Chuck fondly recalls his father's devotion to imparting that sense of joy to him and his siblings: "He has this ability to get you to see things and to be overwhelmed by how beautiful they are, to stir the pot in you of joy and amazement and excitement. I'm talking about little unbelievable windows he was at the center of, you felt so happy to be alive, in his presence, in this event that he created. We were on vacation in Maine once. It wasn't great weather, I recall, and we kids were just hanging out, the way kids do, bored. Somehow my father got a little ball and spontaneously created this strange little game, and in a minute some of the other beach kids from next door came over, and suddenly there were four, five, six of us throwing, catching, and laughing. I remember being so utterly excited and joyful. If I look back and think about what's motivated me in my life, having those moments of pure joy, having a vision of how good life can be, a sense of what life can hold—I've gotten that from my father."

Walter used to organize his children to play a game in the winter, testing the aerodynamic quality of paper airplanes—by flying them into the family's big open living room fireplace. "To my mother's horror," Chuck recalled, "we would recover them from the fire—we were determined to win the competition the next time round!"

When guests came for dinner, Walter would preside over the game of Going to the Moon. As Chuck remembers it, "We would dim the lights, pound our fists on the table making a drumroll kind of sound, simulating the noise of a rocket launch. Some of the kids would even go under the table and pound. Then, as we reached space, we stopped the pounding, and once we landed on the Moon, all of us would walk around the living room pretending to be in very low gravity, taking crazy exaggerated steps. Meanwhile, the guests must have been thinking, 'These people are nuts!' But for us kids, it was fantastic! Going to the Moon!"

Walter Lewin has been taking students to the Moon since he first walked into a classroom more than a half century ago. Perpetually entranced by the mystery and beauty of the natural world—from rainbows to neutron stars, from the femur of a mouse to the sounds of music—and by the efforts of scientists and artists to explain, interpret, and represent this world, Walter Lewin is one of the most passionate, devoted, and skillful scientific guides to that world now alive. In the chapters that follow you will be able to experience that passion, devotion, and skill as he uncovers his lifelong love of physics and shares it with you. Enjoy the journey!

—Warren Goldstein

CHAPTER 1

From the Nucleus to Deep Space

It's amazing, really. My mother's father was illiterate, a custodian. Two generations later I'm a full professor at MIT. I owe a lot to the Dutch educational system. I went to graduate school at the Delft University of Technology in the Netherlands, and killed three birds with one stone.

Right from the start, I began teaching physics. To pay for school I had to take out a loan from the Dutch government, and if I taught full time, at least twenty hours a week, each year the government would forgive one-fifth of my loan. Another advantage of teaching was that I wouldn't have to serve in the army. The military would have been the worst, an absolute *disaster* for me. I'm allergic to all forms of authority—it's just in my personality—and I knew I would have ended up mouthing off and scrubbing floors. So I taught math and physics full time, twenty-two contact hours per week, at the Libanon Lyceum in Rotterdam, to sixteen- and seventeen-year-olds. I avoided the army, did not have to pay back my loan, and was getting my PhD, all at the same time.

I also learned to teach. For me, teaching high school students, being able to change the minds of young people in a positive way, that was thrilling. I always tried to make classes interesting but also fun for the students, even though the school itself was quite strict. The classroom doors had transom windows at the top, and one of the headmasters would sometimes climb up on a chair and spy on teachers through the transom. Can you believe it?

I wasn't caught up in the school culture, and being in graduate school, I was boiling over with enthusiasm. My goal was to impart that enthusiasm to my students, to help them see the beauty of the world all around them in a new way, to change them so that they would see the world of physics as beautiful, and would understand that physics is everywhere, that it permeates our lives. What counts, I found, is not what you *cover*, but what you *uncover*. Covering subjects in a class can be a boring exercise, and students feel it. Uncovering the laws of physics and making them see through the equations, on the other hand, demonstrates the process of discovery, with all its newness and excitement, and students love being part of it.

I got to do this also in a different way far outside the classroom. Every year the school sponsored a week-long vacation when a teacher would take the kids on a trip to a fairly remote and primitive campsite. My wife, Huibertha, and I did it once and loved it. We all cooked together and slept in tents. Then, since we were so far from city lights, we woke all the kids up in the middle of one night, gave them hot chocolate, and took them out to look at the stars. We identified constellations and planets and they got to see the Milky Way in its full glory.

I wasn't studying or even teaching astrophysics—in fact, I was designing experiments to detect some of the smallest particles in the universe—but I'd always been fascinated by astronomy. The truth is that just about every physicist who walks the Earth has a love for astronomy. Many physicists I know built their own telescopes when they were in high school. My longtime friend and MIT colleague George Clark ground and polished a 6-inch mirror for a telescope when he was in high school. Why do physicists love astronomy so much? For one thing, many advances in physics—theories of orbital motion, for instance—have resulted from astronomical questions, observations, and theories. But also, astronomy *is* physics, writ large across the night sky: eclipses, comets, shooting stars, globular clusters, neutron stars, gamma-ray bursts, jets, planetary nebulae, supernovae, clusters of galaxies, black holes.

Just look up in the sky and ask yourself some obvious questions: Why is the sky blue, why are sunsets red, why are clouds white? Physics has the answers! The light of the Sun is composed of all the colors of the rainbow. But as it makes its way through the atmosphere it scatters in all directions off air molecules and very tiny dust particles (much smaller than a micron, which is 1/25,000 of an inch). This is called Rayleigh scattering. Blue light scatters the most of all colors, about five times more than red light. Thus when you look at the sky during the day in any direction*, blue dominates, which is why the sky is blue. If you look at the sky from the surface of the Moon (you may have seen pictures), the sky is not blue—it's black, like our sky at night. Why? Because the Moon has no atmosphere.

Why are sunsets red? For exactly the same reason that the sky is blue. When the Sun is at the horizon, its rays have to travel through more atmosphere, and the green, blue, and violet light get scattered the most—filtered out of the light, basically. By the time the light reaches our eyes—and the clouds above us—it's made up largely of yellow, orange, and especially red. That's why the sky sometimes almost appears to be on fire at sunset and sunrise.

Why are clouds white? The water drops in clouds are much larger than the tiny particles that make our sky blue, and when light scatters off these much larger particles, all the colors in it scatter equally. This causes the light to stay white. But if a cloud is very thick with moisture, or if it is in the shadow of another cloud, then not much light will get through, and the cloud will turn dark.

One of the demonstrations I love to do is to create a patch of "blue sky" in my classes. I turn all the lights off and aim a very bright spotlight of white light at the ceiling of the classroom near my blackboard. The

---

*Be careful—never look at the Sun.

spotlight is carefully shielded. Then I light a few cigarettes and hold them in the light beam. The smoke particles are small enough to produce Rayleigh scattering, and because blue light scatters the most, the students see blue smoke. I then carry this demonstration one step further. I inhale the smoke and keep it in my lungs for a minute or so—this is not always easy, but science occasionally requires sacrifices. I then let go and exhale the smoke into the light beam. The students now see white smoke—I have created a white cloud! The tiny smoke particles have grown in my lungs, as there is a lot of water vapor there. So now all the colors scatter equally, and the scattered light is white. The color change from blue light to white light is truly amazing!

With this demonstration, I'm able to answer two questions at once: Why is the sky blue, and why are clouds white? Actually, there is also a third very interesting question, having to do with the polarization of light. I'll get to this in chapter 5.

Out in the country with my students I could show them the Andromeda galaxy, the only one you can see with the naked eye, around 2.5 million light-years away (15 million trillion miles), which is next door as far as astronomical distances go. It's made up of about 200 billion stars. Imagine that—200 billion stars, and we could just make it out as a faint fuzzy patch. We also spotted lots of meteorites—most people call them shooting stars. If you were patient, you'd see one about every four or five minutes. In those days there were no satellites, but now you'd see a host of those as well. There are more than two thousand now orbiting Earth, and if you can hold your gaze for five minutes you'll almost surely see one, especially within a few hours after sunset or before sunrise, when the Sun hasn't yet set or risen on the satellite itself and sunlight still reflects off it to your eyes. The more distant the satellite, and therefore the greater the difference in time between sunset on Earth and at the satellite, the later you can see it at night. You recognize satellites because they move faster than anything else in the sky (except meteorites); if it blinks, believe me, it's an airplane.

I have always especially liked to point out Mercury to people when we're stargazing. As the planet closest to the Sun, it's very difficult to see it with the naked eye. The conditions are best only about two dozen evenings and mornings a year. Mercury orbits the Sun in just eighty-eight days, which is why it was named for the fleet-footed Roman messenger god; and the reason it's so hard to see is that its orbit is so close to the Sun. It's never more than about 25 degrees away from the Sun when we look at it from Earth—that's smaller than the angle between the two hands of a watch at eleven o'clock. You can only see it shortly after sunset and before sunrise, and when it's farthest from the Sun as seen from Earth. In the United States it's always close to the horizon; you almost have to be in the countryside to see it. How wonderful it is when you actually find it!

Stargazing connects us to the vastness of the universe. If we keep staring up at the night sky, and let our eyes adjust long enough, we can see the superstructure of the farther reaches of our own Milky Way galaxy quite beautifully—some 100 billion to 200 billion stars, clustered as if woven into a diaphanous fabric, so delightfully delicate. The size of the universe is incomprehensible, but you can begin to grasp it by first considering the Milky Way.

Our current estimate is that there may be as many galaxies in the universe as there are stars in our own galaxy. In fact, whenever a telescope observes deep space, what it sees is mostly galaxies—it's impossible to distinguish single stars at truly great distances—and each contains billions of stars. Or consider the recent discovery of the single largest structure in the known universe, the Great Wall of galaxies, mapped by the Sloan Digital Sky Survey, a major project that has combined the efforts of more than three hundred astronomers and engineers and twenty-five universities and research institutions. The dedicated Sloan telescope is observing every night; it went into operation in the year 2000 and will continue till at least the year 2014. The Great Wall is more than a billion light-years long. Is your head spinning? If not, then consider that the observable universe (not the entire universe, just the part we can observe) is roughly 90 *billion light-years* across.

This is the power of physics; it can tell us that our observable universe is made up of some 100 billion galaxies. It can also tell us that of all the matter in our visible universe, only about 4 percent is ordinary matter, of which stars and galaxies (and you and I) are made. About 23 percent is what's called dark matter (it's invisible). We know it exists, but we don't know what it is. The remaining 73 percent, which is the bulk of the energy in our universe, is called dark energy, which is also invisible. No one has a clue what that is either. The bottom line is that we're ignorant about 96 percent of the mass/energy in our universe. Physics has explained so much, but we still have many mysteries to solve, which I find very inspiring.

Physics explores unimaginable immensity, but at the same time it can dig down into the very smallest realms, to the very bits of matter such as neutrinos, as small as a tiny fraction of a proton. That is where I was spending most of my time in my early days in the field, in the realms of the very small, measuring and mapping the release of particles and radiation from radioactive nuclei. This was nuclear physics, but not the bomb-making variety. I was studying what made matter tick at a really basic level.

You probably know that almost all the matter you can see and touch is made up of elements, such as hydrogen, oxygen, and carbon combined into molecules, and that the smallest unit of an element is an atom, made up of a nucleus and electrons. A nucleus, recall, consists of protons and neutrons. The lightest and most plentiful element in the universe, hydrogen, has one proton and one electron. But there is a form of hydrogen that has a neutron as well as a proton in its nucleus. That is an isotope of hydrogen, a different form of the same element; it's called deuterium. There's even a third isotope of hydrogen, with two neutrons joining the proton in the nucleus; that's called tritium. All isotopes of a given element have the same number of protons, but a different number of neutrons, and elements have different numbers of isotopes. There are thirteen isotopes of oxygen, for instance, and thirty-six isotopes of gold.

Now, many of these isotopes are stable—that is, they can last more or less forever. But most are unstable, which is another way of saying they're radioactive, and radioactive isotopes decay: that is to say, sooner or later they transform themselves into other elements. Some of the elements they transform into are stable, and then the radioactive decay stops, but others are unstable, and then the decay continues until a stable state is reached. Of the three isotopes of hydrogen, only one, tritium, is radioactive—it decays into a stable isotope of helium. Of the thirteen isotopes of oxygen, three are stable; of gold's thirty-six isotopes, only one is stable.

You will probably remember that we measure how quickly radioactive isotopes decay by their "half-life"—which can range from a microsecond (one-millionth of a second) to billions of years. If we say that tritium has a half-life of about twelve years, we mean that in a given sample of tritium, half of the isotopes will decay in twelve years (only one-quarter will remain after twenty-four years). Nuclear decay is one of the most important processes by which many different elements are transformed and created. It's not alchemy. In fact, during my PhD research, I was often watching radioactive gold isotopes decay into mercury rather than the other way around, as the medieval alchemists would have liked. There are, however, many isotopes of mercury, and also of platinum, that decay into gold. But only one platinum isotope and only one mercury isotope decay into stable gold, the kind you can wear on your finger.

The work was immensely exciting; I would have radioactive isotopes literally decaying in my hands. And it was very intense. The isotopes I was working with typically had half-lives of only a day or a few days. Gold-198, for instance, has a half-life of a little over two and a half days, so I had to work fast. I would drive from Delft to Amsterdam, where they used a cyclotron to make these isotopes, and rush back to the lab at Delft. There I would dissolve the isotopes in an acid to get them into liquid form, put them on very thin film, and place them into detectors.

I was trying to verify a theory about nuclear decay, one that predicted the ratio of gamma ray to electron emissions from the nuclei, and my work required precise measurements. This work had already been done for many radioactive isotopes, but some recent measurements had come out that were different from what the theory predicted. My supervisor,

Professor Aaldert Wapstra, suggested I try to determine whether it was the theory or the measurements that were at fault. It was enormously satisfying, like working on a fantastically intricate puzzle. The challenge was that my measurements had to be much more precise than the ones those other researchers had come up with before me.

Electrons are so small that some say they have no effective size—they're less than a thousand-trillionth of a centimeter across—and gamma rays have a wavelength of less than a billionth of a centimeter. And yet physics had provided me with the means to detect and to count them. That's yet another thing that I love about experimental physics; it lets us "touch" the invisible.

To get the measurements I needed, I had to milk the sample as long as I could, because the more counts I had, the greater my precision would be. I'd frequently be working for something like 60 hours straight, often without sleeping. I became a little obsessed.

For an experimental physicist, precision is key in everything. The accuracy is the *only* thing that matters, and a measurement that doesn't also indicate its degree of accuracy is meaningless. This simple, powerful, totally fundamental idea is almost always ignored in college books about physics. Knowing degrees of accuracy is critical to so many things in our lives.

In my work with radioactive isotopes, attaining the degree of accuracy I had to achieve was very challenging, but over three or four years I got better and better at the measurements. After I improved some of the detectors, they turned out to be extremely accurate. I was confirming the theory, and publishing my results, and this work ended up being my PhD thesis. What was especially satisfying to me was that my results were rather conclusive, which doesn't happen very often. Many times in physics, and in science generally, results are not always clear-cut. I was fortunate to arrive at a firm conclusion. I had solved a puzzle and established myself as a physicist, and I had helped to chart the unknown territory of the subatomic world. I was twenty-nine years old, and I was thrilled to be making a solid contribution. Not all of us are destined to make gigantic fundamental discoveries like Newton and Einstein did, but there's an awful lot of territory that is still ripe for exploration.

I was also fortunate that at the time I got my degree, a whole new era of discovery about the nature of the universe was getting under way. Astronomers were making discoveries at an amazing pace. Some were examining the atmospheres of Mars and Venus, searching for water vapor. Some had discovered the belts of charged particles circling the Earth's magnetic field lines, which we now call the Van Allen belts. Others had discovered huge, powerful sources of radio waves known as quasars (quasi-stellar radio sources). The cosmic microwave background (CMB) radiation was discovered in 1965—the traces of the energy released by the big bang, powerful evidence for the big bang theory of the universe's origin, which had been controversial. Shortly after, in 1967, astronomers would discover a new category of stars, which came to be called pulsars.

I might have continued working in nuclear physics, as there was a great deal of discovery going on there as well. This work was mostly in the hunt for and discovery of a rapidly growing zoo of subatomic particles, most importantly those called quarks, which turned out to be the building blocks of protons and neutrons. Quarks are so odd in their range of behaviors that in order to classify them, physicists assigned them what they called flavors: up, down, strange, charm, top, and bottom. The discovery of quarks was one of those beautiful moments in science when a purely theoretical idea is confirmed. Theorists had predicted quarks, and then experimentalists managed to find them. And how exotic they were, revealing that matter was so much more complicated in its foundations than we had known. For instance, we now know that protons consist of two up quarks and one down quark, held together by the strong nuclear force, in the form of other strange particles called gluons. Some theoreticians have recently calculated that the up quark seems to have a mass of about 0.2 percent of that of a proton, while the down quark has a mass of about 0.5 percent of the mass of a proton. This was not your grandfather's nucleus anymore.

The particle zoo would have been a fascinating area of research to go into, I'm sure, but by a happy accident, the skills I'd learned for measuring radiation emitted from the nucleus turned out to be extremely useful for probing the universe. In 1965, I received an invitation from Professor Bruno Rossi at MIT to work on X-ray astronomy, which was an entirely new field, really just a few years old at the time—Rossi had initiated it in 1959.

MIT was the best thing that could ever have happened to me. Rossi's work on cosmic rays was already legendary. He'd headed a department at Los Alamos during the war and pioneered in the measurements of solar wind, also called interplanetary plasma—a stream of charged particles ejected by the Sun that causes our aurora borealis and "blows" comet tails away from the Sun. Now he had the idea to search the cosmos for X-rays. It was completely exploratory work; he had no idea whether he'd find them or not.

Anything went at that time at MIT. Any idea you had, if you could convince people that it was doable, you could work on it. What a difference from the Netherlands! At Delft, there was a rigid hierarchy, and the graduate students were treated like a lower class. The professors were given keys to the front door of my building, but as a graduate student you only got a key to the door in the basement, where the bicycles were kept. Each time you entered the building you had to pick your way through the bicycle storage rooms and be reminded of the fact that you were *nothing*.

If you wanted to work after five o'clock you had to fill out a form, every day, by four p.m., justifying why you had to stay late, which I had to do almost all the time. The bureaucracy was a real nuisance.

The three professors in charge of my institute had reserved parking places close to the front door. One of them, my own supervisor, worked in Amsterdam and came to Delft only once a week on Tuesdays. I asked him one day, "When you are not here, would you mind if I used your parking space?" He said, "Of course not," but then the very first day I parked there I was called on the public intercom and instructed in the strongest terms possible that I was to remove my car. Here's another one. Since I had to go to Amsterdam to pick up my isotopes, I was allowed 25 cents for a cup of coffee, and 1.25 guilders for lunch (1.25 guilders was about one-third of a U.S. dollar at the time), but I had to submit separate receipts for each. So I asked if I could add the 25 cents to the lunch receipt and only submit one receipt for 1.50 guilders. The department chair, Professor Blaisse, wrote me a letter that stated that if I wanted to have gourmet meals I could do so—at my own expense.

So what a joy it was to get to MIT and be free from all of that; I felt reborn. Everything was done to encourage you. I got a key to the front door and could work in my office day or night just as I wanted. To me, that key to the building was like a key to everything. The head of the Physics Department offered me a faculty position six months after my arrival, in June of 1966. I accepted and I've never left.

Arriving at MIT was also so exhilarating because I had lived through the devastation of World War II. The Nazis had murdered half of my family, a tragedy that I haven't really digested yet. I do talk about it sometimes, but very rarely because it's so very difficult for me—it is more than sixty-five years ago, and it's still overwhelming. When my sister Bea and I talk about it, we almost always cry.

I was born in 1936, and I was just four years old when the Germans attacked the Netherlands on May 10, 1940. One of my earliest memories is all of us, my mother's parents, my mother and father and sister and I, hiding in the bathroom of our house (at the Amandelstraat 61 in The Hague) as the Nazi troops entered my country. We were holding wet handkerchiefs over our noses, as there had been warnings that there would be gas attacks.

The Dutch police snatched my Jewish grandparents, Gustav Lewin and Emma Lewin Gottfeld, from their house in 1942. At about the same time they hauled out my father's sister Julia, her husband Jacob (called Jenno), and her three children—Otto, Rudi, and Emmie—and put them all on trucks, with their suitcases, and sent them to Westerbork, the transshipment camp in Holland. More than a hundred thousand Jews passed through Westerbork, on their way to other camps. The Nazis quickly sent my grandparents to Auschwitz and murdered them—gassed them—the day they arrived, November 19, 1942. My grandfather was seventy-five and my grandmother sixty-nine, so they wouldn't have been candidates for labor camps. Westerbork, by contrast, was so strange; it was made to look like a resort for Jews. There were ballet performances and shops. My mother would often bake potato pancakes that she would then send by mail to our family in Westerbork.

Because my uncle Jenno was what the Dutch call "*statenloos,*" or stateless—he had no nationality—he was able to drag his feet and stay at Westerbork with his family for almost a year before the Nazis split up the family and shipped them to different camps. They sent my aunt Julia and my cousins Emmie and Rudi first to the women's concentration camp Ravensbrück in Germany and then to Bergen-Belsen, also in Germany, where they were imprisoned until the war ended. My aunt Julia died ten days after the camp's liberation by the Allies, but my cousins survived. My cousin Otto, the oldest, had also been sent to Ravensbrück, to the men's camp there, and near the end of the war ended up in the concentration camp in Sachsenhausen; he survived the Sachsenhausen death march in April 1945. Uncle Jenno they sent directly to Buchenwald, where they murdered him—along with more than 55,000 others.

Whenever I see a movie about the Holocaust, which I would not do for a really long time, I project it immediately onto my own family. That's why I felt the movie *Life Is Beautiful* was terribly difficult to watch, even objectionable. I just couldn't imagine joking about something that was so serious. I still have recurring nightmares about being chased by Nazis, and I wake up sometimes absolutely terrified. I even once in my dreams witnessed my own execution by the Nazis.

Some day I would like to take the walk, my paternal grandparents' last walk, from the train station to the gas chambers at Auschwitz. I don't know if I'll ever do it, but it seems to me like one way to memorialize them. Against such a monstrosity, maybe small gestures are all that we have. That, and our refusal to forget: I never talk about my family members having "died" in concentration camps. I always use the word *murdered*, so we do not let language hide the reality.

My father was Jewish but my mother was not, and as a Jew married to a non-Jewish woman, he was not immediately a target. But he became a target soon enough, in 1943. I remember that he had to wear the yellow star. Not my mother, or sister, or I, but he did. We didn't pay much attention to it, at least not at first. He had it hidden a little bit, under his clothes, which was forbidden. What was really frightening was the way he gradually accommodated to the Nazi restrictions, which just kept getting worse. First, he was not allowed on public transportation. Then, he wasn't allowed in public parks. Then he wasn't allowed in restaurants; he became persona non grata in places he had frequented for years! And the incredible thing is the ability of people to adjust.

When he could no longer take public transportation, he would say, "Well, how often do I make use of public transportation?" When he wasn't allowed in public parks anymore, he would say, "Well, how often do I go to public parks?" Then, when he could not go to a restaurant, he would say, "Well, how often do I go to restaurants?" He tried to make these awful things seem trivial, like a minor inconvenience, perhaps for his children's sake, and perhaps also for his own peace of mind. I don't know.

It's still one of the hardest things for me to talk about. Why this ability to slowly see the water rise but not recognize that it will drown you? How could they see it and not see it at the same time? That's something that I cannot cope with. Of course, in a sense it's completely understandable; perhaps that's the only way you can survive, for as long as you are able to fool yourself.

Though the Nazis made public parks off-limits to Jews, my father was allowed to walk in cemeteries. Even now, I recall many walks with him at a nearby cemetery. We fantasized about how and why family members died—sometimes four had died on the same day. I still do that nowadays when I walk in Cambridge's famous Mount Auburn Cemetery.

The most dramatic thing that happened to me growing up was that all of a sudden my father disappeared. I vividly remember the day he left. I came home from school and sensed somehow that he was gone. My mother was not home, so I asked our nanny, Lenie, "Where's Dad?" and I got an answer of some sort, meant to be reassuring, but somehow I knew that my father had left.

Bea saw him leaving, but she never told me until many years later. The four of us slept in the same bedroom for security, and at four in the morning, she saw him get up and put some clothes in a bag. Then he kissed my mother and left. My mother didn't know where he was going; that knowledge would have been very dangerous, because the Germans might have tortured her to find out where my father was and she would have told them. We now know that the Resistance hid him, and eventually we got some messages from him through the Resistance, but at the time it was absolutely terrible not knowing where he was or even if he was alive.

I was too young to understand how profoundly his absence affected my mother. My parents ran a school out of our home—which no doubt had a strong influence on my love of teaching—and she struggled to carry on without him. She had a tendency toward depression anyway, but now her husband was gone, and she worried that we children might be sent to a concentration camp. She must have been truly terrified for us because—as she told me fifty-five years later—one night she said to Bea and me that we should sleep in the kitchen, and she stuffed curtains and blankets and towels under the doors so that no air could escape. She was intending to put the gas on and let us sleep ourselves into death, but she didn't go through with it. Who can blame her for thinking of it—I know that Bea and I don't.

I was afraid a lot. And I know it sounds ridiculous, but I was the only male, so I sort of became the man of the house, even at age seven and eight. In The Hague, where we lived, there were many broken-down houses on the coast, half-destroyed by the Germans who were building bunkers on our beaches. I would go there and steal wood—I was going to say "collect," but it was stealing—from those houses so that we had some fuel for cooking and for heat.

To try to stay warm in the winters we wore this rough, scratchy, poor-quality wool. And I still cannot stand wool to this day. My skin is so sensitive that I sleep on eight-hundred-thread-count cotton sheets. That's also why I order very fine cotton shirts—ones that do not irritate my skin. My daughter Pauline tells me that if I see her wearing wool, I still turn away; such is the effect the war still has on me.

My father returned while the war was still going on, in the fall of 1944. People in my family disagree about just how this happened, but as near as I can tell it seems that my wonderful aunt Lauk, my mother's sister, was in Amsterdam one day, about 30 miles away from The Hague, and she caught sight of my father! She followed him from a distance and saw him go into a house. Later she went back and discovered that he was living with a woman.

My aunt told my mother, who at first got even more depressed and upset, but I'm told that she collected herself and took the boat to Amsterdam (trains were no longer operating), marched right up to the house, and rang the bell. Out came the woman, and my mother said, "I want to speak to my husband." The woman replied, "*I* am the wife of Mr. Lewin." But my mother insisted: "I want my husband." My father came to the door, and she said, "I'll give you five minutes to pack up and come back with me or else you can get a divorce and you'll never see your children again." In three minutes he came back downstairs with his things and returned with her.

In some ways it was much worse when he was back, because people knew that my father, whose name was also Walter Lewin, was a Jew. The Resistance had given him false identification papers, under the name of Jaap Horstman, and my sister and I were instructed to call him Uncle Jaap. It's a total miracle, and doesn't make any sense to Bea and me to this very day, but no one turned him in. A carpenter made a hatch in the ground floor of our house. We could lift it up and my father could go down and hide in the crawl space. Remarkably, my father managed to avoid capture.

He was probably at home eight months or so before the war ended, including the worst time of the war for us, the winter of 1944 famine, the *hongerwinter*. People starved to death—nearly twenty thousand died. For heat we crawled under the house and pulled out every other floor joist—the large beams that supported the ground floor—for firewood. In the hunger winter we ate tulip bulbs, and even bark. People could have turned my father in for food. The Germans would also pay money (I believe it was fifty guilders, which was about fifteen dollars at the time) for every Jew they turned in.

The Germans did come to our house one day. It turned out that they were collecting typewriters, and they looked at ours, the ones we used to teach typing, but they thought they were too old. The Germans in their own way were pretty stupid; if you're being told to collect typewriters, you don't collect Jews. It sounds like a movie, I know. But it really happened.

After all of the trauma of the war, I suppose the amazing thing is that I had a more or less normal childhood. My parents kept running their school—the Haagsch Studiehuis—which they'd done before and during the war, teaching typing, shorthand, languages, and business skills. I too was a teacher there while I was in college.

My parents patronized the arts, and I began to learn about art. I had an academically and socially wonderful time in college. I got married in 1959, started graduate school in January 1960, and my first daughter, Pauline, was born later that year. My son Emanuel (who is now called Chuck) was born two years after that, and our second daughter, Emma, came in 1965. Our second son, Jakob, was born in the United States in 1967.

When I arrived at MIT, luck was on my side; I found myself right in the middle of the explosion of discoveries going on at that time. The expertise I had to offer was perfect for Bruno Rossi's pioneering X-ray astronomy team, even though I didn't know anything about space research.

V-2 rockets had broken the bounds of the Earth's atmosphere, and a whole new vista of opportunity for discoveries had been opened up.

Ironically, the V-2 had been designed by Wernher von Braun, who was a Nazi. He developed the rockets during World War II to kill Allied civilians, and they were terribly destructive. In Peenemünde and in the notorious underground Mittelwerk plant in Germany, slave laborers from concentration camps built them, and some twenty thousand died in the process. The rockets themselves killed more than seven thousand civilians, mostly in London. There was a launch site about a mile from my mother's parents' house close to The Hague. I recall a sizzling noise as the rockets were being fueled and the roaring noise at launch. In one bombing raid the Allies tried to destroy V-2 equipment, but they missed and killed five hundred Dutch civilians instead. After the war the Americans brought von Braun to the United States and he became a hero. That has always baffled me. He was a war criminal!

For fifteen years von Braun worked with the U.S. Army to build the V-2's descendants, the Redstone and Jupiter missiles, which carried nuclear warheads. In 1960 he joined NASA and directed the Marshall Space Flight Center in Alabama, where he developed the Saturn rockets that sent astronauts to the Moon. Descendants of his rockets launched the field of X-ray astronomy, so while rockets began as weapons, at least they also got used for a great deal of science. In the late 1950s and early 1960s they opened new windows on the world—no, on the universe!—giving us the chance to peek outside of the Earth's atmosphere and look around for things we couldn't see otherwise.

To discover X-rays from outer space, Rossi had played a hunch. In 1959 he went to an ex-student of his named Martin Annis, who then headed a research firm in Cambridge called American Science and Engineering, and said, "Let's just see if there are X-rays out there." The ASE team, headed by future Nobelist Riccardo Giacconi, put three Geiger-Müller counters in a rocket that they launched on June 18, 1962. It spent just six minutes above 80 kilometers (about 50 miles), to get beyond the Earth's atmosphere—a necessity, since the atmosphere absorbs X-rays.

Sure enough, they detected X-rays, and even more important, they were able to establish that the X-rays came from a source outside the solar system. It was a bombshell that changed all of astronomy. No one expected it, and no one could think of plausible reasons why they were there; no one really understood the finding. Rossi had been throwing an idea at the wall to see if it would stick. These are the kinds of hunches that make a great scientist.

I remember the exact date I arrived at MIT, January 11, 1966, because one of our kids got the mumps and we had to delay going to Boston; the KLM wouldn't let us fly, as the mumps is contagious. On my first day I met Bruno Rossi and also George Clark, who in 1964 had been the first to fly a balloon at a very high altitude—about 140,000 feet—to search for X-ray sources that emitted very high energy X-rays, the kind that could penetrate down to that altitude. George said, "If you want to join my group that would be great." I was at exactly the right place at the right time.

If you're the first to do something, you're bound to be successful, and our team made one discovery after another. George was very generous; after two years he turned the group completely over to me. To be on the cutting edge of the newest wave in astrophysics was just remarkable.

I was incredibly fortunate to find myself right in the thick of the most exciting work going on in astrophysics at that time, but the truth is that all areas of physics are amazing; all are filled with intriguing delights and are revealing astonishing new discoveries all the time. While we were finding new X-ray sources, particle physicists were finding ever more fundamental building blocks of the nucleus, solving the mystery of what holds nuclei together, discovering the W and Z bosons, which carry the "weak" nuclear interactions, and quarks and gluons, which carry the "strong" interactions.

Physics has allowed us to see far back in time, to the very edges of the universe, and to make the astonishing image known as the Hubble Ultra Deep Field, revealing what seems an infinity of galaxies. You should not finish this chapter without looking up the Ultra Deep Field online. I have friends who've made this image their screen saver!

The universe is about 13.7 billion years old. However, due to the fact that space itself has expanded enormously since the big bang, we are currently observing galaxies that were formed some 400 to 800 million years after the big bang and that are now considerably farther away than 13.7 billion light-years. Astronomers now estimate that the edge of the observable universe is about 47 billion light-years away from us in every direction. Because of the expansion of space, many faraway galaxies are currently moving away from us faster than the speed of light. This may sound shocking, even impossible, to those of you raised on the notion that, as Einstein postulated in his theory of special relativity, nothing can go faster than the speed of light. However, according to Einstein's theory of general relativity, there are no limits on the speed between two galaxies when space itself is expanding. There are good reasons why scientists now think that we are living in the golden age of cosmology—the study of the origin and evolution of the entire universe.

Physics has explained the beauty and fragility of rainbows, the existence of black holes, why the planets move the way they do, what goes on when a star explodes, why a spinning ice skater speeds up when she draws in her arms, why astronauts are weightless in space, how elements were formed in the universe, when our universe began, how a flute makes music, how we generate electricity that drives our bodies as well as our economy, and what the big bang sounded like. It has charted the smallest reaches of subatomic space and the farthest reaches of the universe.

My friend and colleague Victor Weisskopf, who was already an elder statesman when I arrived at MIT, wrote a book called *The Privilege of Being a Physicist*. That wonderful title captures the feelings I've had being smack in the middle of one of the most exciting periods of astronomical and astrophysical discovery since men and women started looking carefully at the night sky. The people I've worked alongside at MIT, sometimes right across the hall from me, have devised astonishingly creative and sophisticated techniques to hammer away at the most fundamental questions in all of science. And it's been my own privilege both to help extend humankind's collective knowledge of the stars and the universe and to bring several generations of young people to an appreciation and love for this magnificent field.

Ever since those early days of holding decaying isotopes in the palm of my hand, I have never ceased to be delighted by the discoveries of physics, both old and new; by its rich history and ever-moving frontiers; and by the way it has opened my eyes to unexpected wonders of the world all around me. For me physics is a way of seeing—the spectacular and the mundane, the immense and the minute—as a beautiful, thrillingly interwoven whole.

That is the way I've always tried to make physics come alive for my students. I believe it's much more important for them to remember the beauty of the discoveries than to focus on the complicated math—after all, most of them aren't going to become physicists. I have done my utmost to help them see the world in a different way; to ask questions they've never thought to ask before; to allow them to see rainbows in a way they have never seen before; and to focus on the exquisite beauty of physics, rather than on the minutiae of the mathematics. That is also the intention of this book, to help open your eyes to the remarkable ways in which physics illuminates the workings of our world and its astonishing elegance and beauty.

CHAPTER 2

Measurements, Uncertainties, and the Stars

My Grandmother and Galileo Galilei

Physics is fundamentally an experimental science, and measurements and their uncertainties are at the heart of every experiment, every discovery. Even the great theoretical breakthroughs in physics come in the form of predictions about quantities that can be measured. Take, for example, Newton's second law, $F = ma$ (force equals mass times acceleration), perhaps the most important single equation in physics, or Einstein's $E = mc^2$ (energy equals mass times the square of the speed of light), the most renowned equation in physics. How else do physicists express relationships except through mathematical equations about measurable quantities such as density, weight, length, charge, gravitational attraction, temperature, or velocity?

I will admit that I may be a bit biased here, since my PhD research consisted of measuring different kinds of nuclear decay to a high degree of accuracy, and that my contributions in the early years of X-ray astronomy came from my measurements of high-energy X-rays from tens of thousands of light-years away. But there simply is no physics without measurements. And just as important, there are no meaningful measurements without their uncertainties.

You count on reasonable amounts of uncertainty all the time, without realizing it. When your bank reports how much money you have in your account, you expect an uncertainty of less than half a penny. When you buy a piece of clothing online, you expect its fit not to vary more than a very small fraction of a size. A pair of size 34 pants that varies just 3 percent changes a full inch in waist size; it could end up a 35 and hang on your hips, or a 33 and make you wonder how you gained all that weight.

It's also vital that measurements are expressed in the right units. Take the case of an eleven-year-long mission costing $125 million—the Mars Climate Orbiter—which came to a catastrophic conclusion because of a confusion in units. One engineering team used metric units while another used English ones, and as a result in September 1999 the spacecraft entered the Martian atmosphere instead of reaching a stable orbit.

In this book I use metric units most of the time because most scientists use them. From time to time, however, I'll use English units—inches, feet, miles, and pounds—when it seems appropriate for a U.S. audience. For temperature, I'll use the Celsius or Kelvin (Celsius plus 273.15) scales but sometimes Fahrenheit, even though no physicist works in degrees Fahrenheit.

My appreciation of the crucial role of measurements in physics is one reason I'm skeptical of theories that can't be verified by means of measurements. Take string theory, or its souped-up cousin superstring theory, the latest effort of theoreticians to come up with a "theory of everything." Theoretical physicists, and there are some brilliant ones doing string theory, have yet to come up with a single experiment, a single prediction that could test any of string theory's propositions. Nothing in string theory can be experimentally verified—at least so far. This means that string theory has no predictive power, which is why some physicists, such as Sheldon Glashow at Harvard, question whether it's even physics at all.

However, string theory has some brilliant and eloquent proponents. Brian Greene is one, and his book and PBS program *The Elegant Universe* (I'm interviewed briefly on it) are charming and beautiful. Edward Witten's M-theory, which unified five different string theories and posits that there are eleven dimensions of space, of which we lower-order beings see only three, is pretty wild stuff and is intriguing to contemplate.

But when theory gets way out there, I am reminded of my grandmother, my mother's mother, a very great lady who had some wonderful sayings and habits that showed her to be quite an intuitive scientist. She used to tell me, for instance, that you are shorter when standing up than when lying down. I love to teach my students about this. On the first day of class I announce to them that in honor of my grandmother, I'm going to bring this outlandish notion to a test. They, of course, are completely bewildered. I can almost see them thinking, "Shorter standing up than lying down? Impossible!"

Their disbelief is understandable. Certainly if there is any difference in length between lying down and standing up it must be quite small. After all, if it was one foot, you'd know it, wouldn't you? You'd get out of bed in the morning, you'd stand up and go *clunk*—you're one foot shorter. But if the difference was only 0.1 centimeters (¹⁄₂₅ of an inch) you might never know. That's why I suspect that if my grandmother was right, then the difference is probably only a few centimeters, maybe as much as an inch.

To conduct my experiment, I of course first need to convince them of the uncertainty in my measurements. So I begin by measuring an aluminum rod vertically—it comes to 150.0 centimeters—and I ask them to agree that I'm probably capable of measuring it with an uncertainty of plus or minus one-tenth of a centimeter. So that vertical measurement is 150.0 ± 0.1 centimeters. I then measure the bar when it's horizontal and come up with 149.9 ± 0.1 centimeters, which is in agreement—within the uncertainty of the measurements—with the vertical measurement.

What did I gain by measuring the aluminum rod in both positions? A lot! For one, the two measurements demonstrate that I was able to measure length to an accuracy of about 0.1 centimeter (1 millimeter). But at least as important for me is the fact that I want to prove to the students that I'm not playing games with them. Suppose, for example, that I have prepared a specially "cooked" meter stick for my horizontal measurements—that would be a terrible, very dishonest thing to do. By showing that the length of the aluminum rod is the same in the two measurements, I establish that my scientific integrity is beyond doubt.

I then ask for a volunteer, measure him standing up, write that number on the blackboard—185.2 centimeters (or just over 6 feet), plus or minus 0.1 centimeter of course, to account for the uncertainty. Then I help him lie down on my desk in my measuring equipment, which looks like a giant Ritz Stick, the wooden shoe-store foot-measuring device, only his whole body is the foot. I joke back and forth with him about how comfortable he is and congratulate him on his sacrifice for the sake of science, which makes him just a wee bit uneasy. What have I got up my sleeve? I slide the triangular wooden block snug up against his head, and while he lies there, I write the new number on the board. So we now have two measurements, each uncertain by about 0.1 centimeters. What's the result?

Are you surprised to learn that the two measurements differ by 2.5 centimeters, plus or minus 0.2 centimeters of course? I have to conclude that he is in fact at least 2.3 centimeters (or about 0.9 inches) taller while lying down. I go back to my prone student, announce that he's roughly an inch taller sleeping than standing up, and—this is the best part—declare, "My grandmother was right! She was always right!"

Are you skeptical? Well, it turns out that my grandmother was a better scientist than most of us. When we are standing, the tug of gravity compresses the soft tissue between the vertebrae of our spines, and when we lie down, our spines expand. This may seem obvious once you know it, but would you have predicted it? In fact, not even the scientists at NASA anticipated this effect in planning the first space missions. The astronauts complained that their suits got tighter when they were in space. Studies done later, during the Skylab mission, showed that of the six astronauts who were measured, all six showed about 3 percent growth in height—a little over 2 inches if you're 6 feet tall. Now astronauts' suits are made with extra room to allow for this growth.

See how revealing good measurements can be? In that same class where I prove my grandmother right, I have a lot of fun measuring some very odd items, all in order to test a suggestion of the great Galileo Galilei, the father of modern science and astronomy, who once asked himself the question, "Why are the largest mammals as large as they are and not much larger?" He answered himself by suggesting that if a mammal became too heavy, its bones would break. When I read about this, I was intrigued to find out whether or not he was right. His answer seemed right intuitively, but I wanted to check it.

I knew that mammals' femurs—their thighbones—support most of their weight, so I decided to make some comparative measurements of different mammals' femur bones. If Galileo was right, then for a super heavy mammal, the femur bone would not be strong enough to support the animal. Of course, I realized that the strength of the mammal's femur should depend on its thickness. Thicker bones can support more weight—that's intuitive. The bigger the animal, the stronger the bones would need to be.

The femur would also get longer as the animal got bigger, of course, and I realized that by comparing how much longer versus how much thicker the femurs of various mammals get as the animals become bigger, I could test Galileo's idea. According to the calculations I made, which are more complicated than I want to go into here (I explain them in appendix 1), I determined that if Galileo was right, then as mammals get bigger the thickness of their femurs would have to increase faster than their length. I calculated that, for example, if one animal was five times bigger than another—so the femur would be five times longer—then the thickness of its femur would have to be about eleven times greater.

This would mean that at some point the thicknesses of femurs would become the same as their lengths—or even greater—which would make for some pretty impractical mammals. Such an animal would certainly not be the fittest for survival, and that would then be the reason why there is a maximum limit on the size of mammals.

So, I had my prediction that thickness would increase faster than length. Now the real fun began.

I went over to Harvard University, where they have a beautiful collection of bones, and I asked them for the femurs of a raccoon and a horse. It turns out that a horse is about four times larger than a raccoon, and sure enough, the horse's femur (42.0 ± 0.5 centimeters) was about three and a half times longer than the raccoon's (12.4 ± 0.3 centimeters). So far so good. I plugged the numbers into my formula and predicted that the horse's femur should be a little more than six times thicker than the raccoon's. When I measured the thicknesses (to an uncertainty of about half a centimeter for the raccoon and 2 centimeters for the horse), it turned out that the horse bone was five times thicker, plus or minus about 10 percent. So it looked very good for Galileo. However, I decided to expand the data to include smaller as well as larger mammals.

So I went back to Harvard, and they gave me three more bones, of an antelope, an opossum, and a mouse. Here's how they all stacked up:

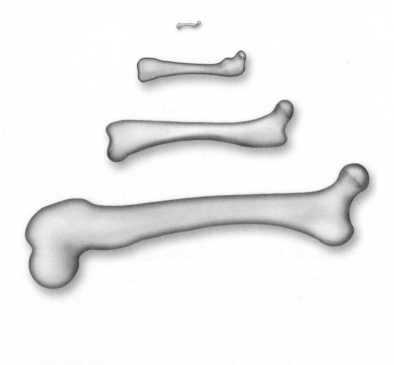

Isn't that wonderful, so romantic? The progression of shapes is lovely, and look at how delicate, how tiny is the femur of the mouse. Only a teeny weenie little femur for a teeny, weenie little mouse. Isn't that beautiful? I will never cease to be amazed by the beauty in every detail of our natural world.

But what about the measurements; how did they fit into my equation? When I did the calculations, I was shocked, really shocked. The horse femur is about 40 times longer than the mouse's, and my calculations predicted that its femur should be more than 250 times thicker. Instead, it was only about 70 times thicker.

So I said to myself, "Why didn't I ask them for the femur of an elephant? That might settle the issue conclusively." I think they were somewhat annoyed at Harvard when I came back again, but they kindly gave me the femur of an elephant. By that time I'm sure they just wanted to get rid of me! Believe me, it was difficult carrying that bone; it was more than a yard long and weighed a ton. I couldn't wait to do my measurements; I couldn't sleep all night.

And do you know what I found? The mouse's femur was 1.1 ± 0.05 centimeters long and 0.7 ± 0.1 millimeters thick—very thin indeed. The elephant's femur was 101 ± 1 centimeters long, about 100 times longer than that of the mouse. So how about its thickness? I measured it at 86 ± 4 millimeters, roughly 120 times the diameter of the mouse's femur. But according to my calculations, if Galileo was right, the femur of the elephant should be roughly 1,000 times thicker than that of the mouse. In other words, it should have been about 70 centimeters thick. Instead, the actual thickness was only about 9 centimeters. I concluded, however reluctantly, that the great Galileo Galilei was wrong!

Measuring Interstellar Space

One of the areas of physics in which measurement has been bedeviling is astronomy. Measurements and uncertainties are enormous issues for astronomers, especially because we deal with such immense distances.

How far away are the stars? How about our beautiful neighbor, the Andromeda Galaxy? And what about all the galaxies we can see with the most powerful telescopes? When we see the most-distant objects in space, how far are we seeing? How large is the universe?

These are some of the most fundamental and profound questions in all of science. And the different answers have turned our view of the universe upside down. In fact, the whole distance business has a wonderful history. You can trace the evolution of astronomy itself through the changing techniques of calculating stellar distances. And at every stage these are dependent on the degree of accuracy of measurements, which is to say the equipment and the inventiveness of astronomers. Until the end of the nineteenth century, the only way astronomers could make these calculations was by measuring something called parallax.

You are all familiar with the phenomenon of parallax without even realizing it. Wherever you are sitting, look around and find a stretch of wall with some sort of feature along it—a doorway or a picture hanging on it—or if you're outside some feature of the landscape, like a big tree. Now stretch your hand straight out in front of you and raise one finger so that it is to one or the other side of that feature. Now first close your right eye and then close your left eye. You will see that your finger jumped from left to right relative to the doorway or the tree. Now, move your finger closer to your eyes and do it again. Your finger moves even more. The effect is *huge*! This is parallax.

It happens because of the switch to different lines of sight in observing an object, so in this case from the line of sight of your left eye to that of your right eye (your eyes are about 6.5 centimeters apart).

This is the basic idea behind determining distances to stars. Except that instead of the approximately 6.5 centimeters separation of my eyes as our baseline, we now use the diameter of the Earth's orbit (about 300 million kilometers) as our baseline. As the Earth moves around the Sun in one year (in an orbit with a diameter of about 300 million kilometers) a nearby star will move in the sky relative to more distant stars. We measure the angle in the sky (called a parallax angle) between the two positions of the star measured six months apart. If you make many sets of measurements all six months apart, you will find different parallax angles. In the figure below, for simplicity, I have selected a star in the same plane of space as Earth, known as the orbital plane (also called the ecliptic plane). However, the principle of parallax measurements as described here holds for any star—not just for stars in the ecliptic plane.

Suppose you observe the star when the Earth is located at position 1 in its orbit around the Sun. You will then see the star projected on the background (very far away) in the direction A1. If now you observe the same star six months later (from position 7), you will see the star in the direction A7. The angle marked as α is the largest possible parallax angle. If you make similar measurements from positions 2 and 8, 3 and 9, 4 and 10, you will then always find parallax angles that are smaller than α. In the hypothetical case of observations from points 4 and 10 (hypothetical, as the star cannot be observed from position 10 since the Sun is then in the way), the parallax angle would even be zero. Now look at the triangle that is formed by the points 1A7. We know that the distance 1–7 is 300 million kilometers, and we know the angle α. Thus we can now calculate the distance *SA* (with high school math).

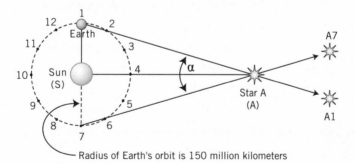

Radius of Earth's orbit is 150 million kilometers

Even though the parallax angles taken at different six-month intervals vary, astronomers talk about *the* parallax of a star. What they mean by that is half the largest parallax angle. If the maximum parallax angle was 2.00 arc seconds, then *the* parallax would be 1.00 arc seconds and the distance to the star would then be 3.26 light-years (however, there is no star that close to us). The smaller *the* parallax, the greater the distance. If *the* parallax is 0.10 arc seconds, its distance is 32.6 light-years. The star nearest the Sun is Proxima Centauri. Its parallax is 0.76 arc seconds; thus its distance is about 4.3 light-years.

To understand just how small the changes in stellar positions are that astronomers must measure, we have to understand just how small an arc second is. Picture an enormous circle drawn in the night sky going through the zenith (which is straight overhead) all the way around the Earth. That circle of course contains 360 degrees. Now each degree is divided into 60 arc minutes, and each arc minute is divided in turn into 60 arc seconds. So there are 1,296,000 arc seconds in that full circle. You can see that an arc second is extremely small.

Here's another way to envision how small. If you take a dime and move it 2.2 miles away from you, its diameter would be one arc second. And here's another. Every astronomer knows that the Moon is about half a degree across, or 30 arc minutes. This is called the angular size of the Moon. If you could cut the Moon into 1,800 equally thin slices, each one would be an arc second wide.

Since the parallax angles that astronomers must measure in order to determine distances are so very small, you may appreciate how important the degree of uncertainty in the measurements is for them.

As improvements in equipment have allowed astronomers to make more and more accurate measurements, their estimates of stellar distances have changed, sometimes quite dramatically. In the early nineteenth century Thomas Henderson measured the parallax of the brightest star in the heavens, Sirius, to be 0.23 arc seconds, with an uncertainty of about a quarter of an arc second. In other words, he had measured an upper limit for the parallax of about half an arc second, and that meant that the star could not be closer to us than 6.5 light-years. In 1839 this was a very important result. But a half century later, David Gill measured Sirius's parallax at 0.370 arc seconds with an uncertainty of plus or minus 0.010 arc seconds. Gill's measurements were consistent with Henderson's, but Gill's measurements were highly superior because the uncertainty was twenty-five times smaller. At a parallax of 0.370 ± 0.010 arc seconds, the distance to Sirius becomes 8.81 ± 0.23 light-years, which is indeed larger than 6.5 light-years!

In the 1990s Hipparcos, the High Precision Parallax Collecting Satellite (I think they fiddled with the name until it matched the name of a famous ancient Greek astronomer), measured the parallaxes of (and hence the distances to) more than a hundred thousand stars with an uncertainty of only about a thousandth of an arc second. Isn't that incredible? Remember how far away that dime had to be to represent an arc second? To cover a thousandth of an arc second, it would have to be 2,200 miles away from an observer.

One of the stars Hipparcos measured the parallax of was, of course, Sirius, and the result was 0.37921 ± 0.00158 arc seconds. This gives a distance to Sirius of 8.601 ± 0.036 light-years.

By far the most accurate parallax measurement ever made was by radio astronomers during the years 1995 to 1998 for a very very special star called Sco X-1. I will tell you all about it in chapter 10. They measured a parallax of 0.00036 ± 0.00004 arc seconds, which translates into a distance of 9.1 ± 0.9 thousand light-years.

In addition to the uncertainties that we must deal with in astronomy as a consequence of the limited accuracy of our equipment, and also to limits in available observation time, there are the astronomers' nightmares: the "unknown-hidden" uncertainties. Is there perhaps an error you are making that you don't even know about because you're missing something, or because your instruments are calibrated incorrectly? Suppose your bathroom scale is set to show zero at 10 pounds and has been that way since you bought it. You only discover the error when you go to the doctor—and nearly have a heart attack. We call that a systematic error, and it scares the hell out of us. I'm no fan of former secretary of defense Donald Rumsfeld, but I did feel a tiny bit of sympathy when he said, in a 2002 press briefing, "We know there are some things we do not know. But there are also unknown unknowns—the ones we don't know we don't know."

The challenges of the limits of our equipment make the achievement of one brilliant but mostly ignored female astronomer, Henrietta Swan

Leavitt, all the more astonishing. Leavitt was working at the Harvard Observatory in a low-level staff position in 1908 when she started this work, which enabled a giant jump in measuring the distance to stars.

This kind of thing has happened so often in the history of science that it should be considered a systematic error: discounting the talent, intellect, and contributions of female scientists.*

Leavitt noticed, in the course of her job analyzing thousands of photographic plates of the Small Magellanic Cloud (SMC), that with a certain class of large pulsating stars (now known as Cepheid variables), there was a relationship between the star's optical brightness and the time it took for one complete pulsation, known as the star's period. She found that the longer the period, the brighter the star. As we will see, this discovery opened the door to accurately measuring distances to star clusters and galaxies.

To appreciate the discovery, we first must understand the difference between brightness and luminosity. Optical brightness is the amount of energy per square meter per second of light we receive on Earth. This is measured using optical telescopes. Optical luminosity, on the other hand, is the amount of energy per second radiated by an astronomical object.

Take Venus, often the brightest object in the entire night sky, even brighter than Sirius, which is the brightest star in the sky. Venus is very close to Earth; it's therefore very bright, but it has virtually no intrinsic luminosity. It radiates relatively little energy by comparison to Sirius, a powerful, nuclear-burning furnace twice as massive as our Sun and about twenty-five times as luminous. Knowing an object's luminosity tells astronomers a great deal about it, but the tricky thing about luminosity was that there was no good way to measure it. Brightness is what you measure because it's what you can see; you can't measure luminosity. To measure luminosity you have to know both the star's brightness and its distance.

---

*It happened to Lise Meitner, who helped discover nuclear fission; Rosalind Franklin, who helped discover the structure of DNA; and to Jocelyn Bell, who discovered pulsars and who should have shared in the 1974 Nobel Prize given to her supervisor, Antony Hewish, for "his decisive role in the discovery of pulsars."

Using a technique called statistical parallax, Ejnar Hertzsprung, in 1913, and Harlow Shapley, in 1918, were able to convert Leavitt's brightness values into luminosities. And by assuming that the luminosity of a Cepheid with a given period in the SMC was the same as that of a Cepheid with the same period elsewhere, they had a way to calculate the luminosity relationship for all Cepheids (even those outside the SMC). I won't elaborate here on this method, as it gets quite technical; the important thing to appreciate is that working out the luminosity-period relation was a milestone in measurements of distances. Once you know a star's luminosity and its brightness, you can calculate its distance.

The range in luminosity, by the way, is substantial. A Cepheid with a period of three days has about a thousand times the Sun's luminosity. When its period is thirty days, its luminosity is about thirteen thousand times greater than the Sun's.

In 1923, the great astronomer Edwin Hubble found Cepheids in the Andromeda Galaxy (also known as M31), from which he calculated its distance at about 1 million light-years, a genuinely shocking result to many astronomers. Many, including Shapley, had argued that our own Milky Way contained the entire universe, including M31, and Hubble demonstrated that in fact it was almost unimaginably distant from us. But wait—if you google the distance to the Andromeda Galaxy, you'll find that it's 2.5 million light-years.

This was a case of unknown unknowns. For all his genius, Hubble had made a systematic error. He had based his calculations on the known luminosity of what later came to be known as Type II Cepheids, when in fact he was observing a kind of Cepheid variable about *four times* more luminous than what he thought he was seeing (these were later named Type I Cepheids). Astronomers only discovered the difference in the 1950s, and overnight they realized that their distance measurements for the previous thirty years were off by a factor of two—a large systematic error that *doubled* the size of the known universe.

In 2004, still using the Cepheid variable method, astronomers measured the distance to the Andromeda Galaxy at 2.51 ± 0.13 million light-years. In 2005 another group measured it by using the eclipsing binary stars method, to get a result of 2.52 ± 0.14 million light-years, about 15 million trillion miles. These two measurements are in excellent agreement with each other. Yet the uncertainty is about 140,000 light-years (about 8×10^{17} miles). And this galaxy is by astronomical standards our next-door neighbor. Imagine the uncertainty we have about the distances of so many other galaxies.

You can see why astronomers are always on the hunt for what are called standard candles—objects with known luminosities. They allow us to estimate distances using a range of ingenious ways of establishing reliable tape measures to the cosmos. And they have been vital in establishing what we call the cosmic distance ladder.

We use parallax to measure distances on the first rung on that ladder. Thanks to Hipparcos's fantastically accurate parallax measurements, we can measure the distances of objects up to several thousand light-years with great precision this way. We take the next step with Cepheids, which allow us to obtain good estimates of the distances of objects up to a hundred million light-years away. For the next rungs astronomers use a number of exotic and complicated methods too technical to go into here, many of which depend on standard candles.

The distance measurements become more and more tricky the farther out we want to measure. This is partly due to the remarkable discovery in 1925 by Edwin Hubble that all galaxies in the universe are moving away from one another. Hubble's discovery, one of the most shocking and significant in all of astronomy, perhaps in all of science in the past century, may only be rivaled by Darwin's discovery of evolution through natural selection.

Hubble saw that the light emitted by galaxies showed a distinct shift toward the less energetic end of the spectrum, the "red" end where wavelengths are longer. This is called redshift. The larger the redshift, the faster a galaxy is moving away from us. We know this effect on Earth with sound as the Doppler effect; it explains why we can tell whether an ambulance is coming toward us or going away from us, since the notes are lower when it's speeding away and higher as it speeds toward us. (I will discuss the Doppler shift in more detail in chapter 13.)

For all the galaxies whose redshifts and distance he could measure, Hubble found that the farther away these objects were, the faster they were moving away. So the universe was expanding. What a monumental discovery! Every galaxy in the universe speeding away from every other galaxy.

This can cause great confusion in the meaning of distance when galaxies are billions of light-years away. Do we mean the distance when the light was emitted (13 billion years ago, for instance) or do we mean the distance we think it is now, since the object has substantially increased its distance from us in those 13 billion years? One astronomer may report that the distance is about 13 billion light-years (this is called the light travel time distance) whereas another may report 29 billion light-years for the same object (this is called the co-moving distance).

Hubble's findings have since become known as Hubble's law: the velocity at which galaxies move away from us is directly proportional to their distance from us. The farther away a galaxy is, the faster it is racing away.

Measuring the velocities of the galaxies was relatively easy; the amount of redshift immediately translates into the speed of the galaxy. However, to get accurate distances was a different matter. That was the hardest part. Remember, Hubble's distance to the Andromeda Nebula was off by a factor of 2.5. He came up with the fairly simple equation $v = H_0 D$, where v is the velocity of a given galaxy, D is the distance of that galaxy from us, and H_0 is a constant, now called Hubble's constant. Hubble estimated the constant to be about 500, measured in units of kilometers per second per megaparsec (1 megaparsec is 3.26 million light-years). The uncertainty in his constant was about 10 percent. Thus, as an example, according to Hubble, if a galaxy is at a distance of 5 megaparsecs, its speed relative to us is about 2,500 kilometers per second (about 1,600 miles per second).

Clearly the universe is expanding fast. But that wasn't all Hubble's discovery revealed. If you really knew the value of Hubble's constant, then you could turn the clock backward in order to calculate the time since the big bang, and thus the age of the universe. Hubble himself estimated that the universe was about 2 billion years old. This calculation was in conflict with the age of the Earth, which geologists were just measuring to be upward of 3 billion years. This bothered Hubble mightily, for good reason. Of course, he was unaware of a number of systematic errors he was making. Not only was he confusing different kinds of Cepheid variables in some cases, but he also mistook clouds of gas in which stars were forming for bright stars in faraway galaxies.

One way of looking at eighty years' worth of progress in measuring stellar distances is to look at the history of Hubble's constant itself. Astronomers have been struggling to nail down the value of Hubble's constant for nearly a century, which has produced not only a seven-fold reduction in the constant, which dramatically increased the size of the universe, but also changed the age of the universe, from Hubble's original 2 billion years to our current estimate of nearly 14 billion years—actually 13.75 ± 0.11 billion years. Now, finally, based on observations in part from the fabulous orbiting telescope bearing Hubble's name, we have a consensus that Hubble's constant is 70.4 ± 1.4 kilometers per second per megaparsec. The uncertainty is only 2 percent—which is incredible!

Just think about it. Parallax measurements, starting in 1838, became the foundation for developing the instruments and mathematical tools to reach billions of light-years to the edge of the observable universe.

For all of our remarkable progress in solving mysteries such as this, there are of course a great many mysteries that remain. We can measure the proportion of dark matter and dark energy in the universe, but we have no idea what they are. We know the age of the universe but still wonder when or if and how it will end. We can make very precise measurements of gravitational attraction, electromagnetism, and of the weak and the strong nuclear forces, but we have no clue if they will ever be combined into one unified theory. Nor do we have any idea what the chances are of other intelligent life existing in our own or some other galaxy. So we have a long way to go. But the wonder is just how many answers the tools of physics have provided, to such a remarkably high degree of accuracy.

CHAPTER 3

Bodies in Motion

Here's something fun to try. Stand on a bathroom scale—not one of those fancy ones at your doctor's office, and not one of those digital glass things you have to tap with your toes to make it turn on, just an everyday bathroom scale. It doesn't matter if you have your shoes on (you don't have to impress anyone), and it doesn't matter what number you see, and whether you like it or not. Now, quickly raise yourself up on your toes; then stop and hold yourself there. You'll see that the scale goes a little crazy. You may have to do this several times to clearly see what's going on because it all happens pretty quickly.

First the needle goes up, right? Then it goes way down before it comes back to your weight, where it was before you moved, though depending on your scale, the needle (or numbered disk) might still jiggle a bit before it stabilizes. Then, as you bring your heels down, especially if you do so quickly, the needle first goes down, then shoots up past your weight, before coming to rest back at the weight you may or may not have wanted to know. What was that all about? After all, you weigh the same whether you move your heels down or up on your toes, right? Or do you?

To figure this out, we need, believe it or not, Sir Isaac Newton, my candidate for the greatest physicist of all time. Some of my colleagues disagree, and you can certainly make a case for Albert Einstein, but no one really questions whether Einstein and Newton are the top two. Why do I vote for Newton? Because his discoveries were both so fundamental and so diverse. He studied the nature of light and developed a theory of color. To study the planetary motions he built the first reflecting telescope, which was a major advance over the refracting telescopes of his day, and even today almost all the major telescopes follow the basic principles of his design. In studying the properties of the motion of fluids, he pioneered a major area of physics, and he managed to calculate the speed of sound (he was only off by about 15 percent). Newton even invented a whole new branch of mathematics: calculus. Fortunately, we don't need to resort to calculus to appreciate his most masterful achievements, which have come to be known as Newton's laws. I hope that in this chapter I can show you how far-reaching these apparently simple laws really are.

Newton's Three Laws of Motion

The first law holds that a body at rest will persist in its state of being at rest, and a body in motion will persist in its motion in the same direction with the same speed—unless, in either case, a force acts on it. Or, in Newton's own words, "A body at rest perseveres in its state of rest, or of uniform motion in a right line unless it is compelled to change that state by forces impressed upon it." This is the law of inertia.

The concept of inertia is familiar to us, but if you reflect on it for a bit, you can appreciate how counterintuitive it actually is. We take this law for granted now, even though it runs clearly against our daily experience. After all, things that move rarely do so along a straight line. And they certainly don't usually keep moving indefinitely. We expect them to come to a stop at some point. No golfer could have come up with the law of inertia, since so few putts go in a straight line and so many stop well short of the hole. What was and still is intuitive is the contrary idea—that things naturally tend toward rest—which is why it had dominated Western thinking about these matters for thousands of years until Newton's breakthrough.

Newton turned our understanding of the motion of objects on its head, explaining that the reason a golf ball often stops short of the hole is that the force of friction is slowing it down, and the reason the Moon doesn't shoot off into space, but keeps circling Earth, is that the force of gravitational attraction is holding it in orbit.

To appreciate the reality of inertia more intuitively, think about how difficult it can be when you are ice skating to make the turn at the end of the rink—your body wants to keep going straight and you have to learn just how much force to apply to your skates at just the right angle to move yourself off of that course without flailing wildly or crashing into the wall. Or if you are a skier, think of how difficult it can be to change course quickly to avoid another skier hurtling into your path. The reason we notice inertia so much more in these cases than we generally do is that in both cases there is so little friction acting to slow us down and help us change our motion. Just imagine if putting greens were made of ice; then you would become acutely aware of just how much the golf ball wants to keep going and going.

Consider just how revolutionary an insight this was. Not only did it overturn all previous understanding; it pointed the way to the discovery of a host of forces that are acting on us all the time but are invisible—like friction, gravity, and the magnetic and electric forces. So important was his contribution that in physics the unit of force is called a newton. But not only did Newton allow us to "see" these hidden forces; he also showed us how to measure them.

With the second law he provided a remarkably simple but powerful guide for calculating forces. Considered by some the most important equation in all of physics, the second law is the famous $F = ma$. In words: the *net* force, F, on an object is the mass of the object, m, multiplied by the *net* acceleration, a, of the object.

To see just one way in which this formula is so useful in our daily lives, take the case of an X-ray machine. Figuring out how to produce just the right range of energies for the X-rays is crucial. Here's how Newton's equation lets us do just that.

One of the major findings in physics—which we'll explore more later—is that a charged particle (say an electron or proton or ion) will experience a force when it is placed in an electric field. If we know the charge of the particle and the strength of the electric field, we can calculate the electric force acting on that particle. However, once we do know the force, using Newton's second law we can calculate the acceleration of the particle.*

In an X-ray machine electrons are accelerated before they strike a target inside the X-ray tube. The speed with which the electrons hit the target determines the energy range of the X-rays that are then produced. By changing the strength of the electric field, we can change the acceleration of the electrons. Thus the speed with which the electrons hit the target can be controlled to select the desired energy range of the X-rays.

In order to facilitate making such calculations, physicists use as a unit of force, the newton—1 newton is the force that accelerates a mass of 1 kilogram at 1 meter per second per second. Why do we say "per second per second"? Because with acceleration, the velocity is constantly changing; so, in other words, it doesn't stop after the first second. If the acceleration is constant, the velocity is changing by the same amount every second.

To see this more clearly, take the case of a bowling ball dropped from a tall building in Manhattan—why not from the observation deck of the Empire State Building? It is known that the acceleration of objects dropped on Earth is approximately 9.8 meters per second per second; it is called the gravitational acceleration, represented in physics by g. (For

---

*I have assumed here that the force on the charged particle due to gravity is so small that it can be ignored.

simplicity I am ignoring air drag for now; more about this later.) After the first second the bowling ball has a speed of 9.8 meters per second. By the end of the second second, it will pick up an additional 9.8 meters per second of speed, so it will be moving at 19.6 meters per second. And by the end of the third second it will be traveling 29.4 meters per second. It takes about 8 seconds for the ball to hit the ground. Its speed is then about 8 times 9.8, which is about 78 meters per second (about 175 miles per hour).

What about the much repeated notion that if you threw a penny off the top of the Empire State Building it would kill someone? I'll again exclude the role of air drag, which I emphasize would be considerable in this case. But even without that factored in, a penny hitting you with a speed of about 175 miles per hour will probably not kill you.

This is a good place to grapple with an issue that will come up over and over in this book, mainly because it comes up over and over in physics: the difference between mass and weight. Note that Newton used mass in his equation rather than weight, and though you might think of the two as being the same, they're actually fundamentally different. We commonly use the pound and the kilogram (the units we'll use in this book) as units of weight, but the truth is that they are units of mass.

The difference is actually simple. Your mass is the same no matter where you are in the universe. That's right—on the Moon, in outer space, or on the surface of an asteroid. It's your *weight* that varies. So what is weight, then? Here's where things get a little tricky. Weight is the result of gravitational attraction. Weight is a force: it is mass times the gravitational acceleration ($F = mg$). So our weight varies depending upon the strength of gravity acting on us, which is why astronauts weigh less on the Moon. The Moon's gravity is about a sixth as strong as Earth's, so on the Moon astronauts weigh about one-sixth what they weigh on Earth.

For a given mass, the gravitational attraction of the Earth is about the same no matter where you are on it. So we can get away with saying, "She

weighs a hundred twenty pounds"* or "He weighs eighty kilograms,"*
even though by doing so we are confusing these two categories (mass
and weight). I thought long and hard about whether to use the technical physics unit for force (thus weight) in this book instead of kilos
and pounds, and decided against it on the grounds that it would be too
confusing—no one, not even a physicist whose mass is 80 kilograms
would say, "I weigh seven hundred eighty-four newtons" ($80 \times 9.8 = 784$).
So instead I'll ask you to remember the distinction—and we'll come back
to it in just a little while, when we return to the mystery of why a scale
goes crazy when we stand on our tiptoes on it.

The fact that gravitational acceleration is effectively the same everywhere on Earth is behind a mystery that you may well have heard of:
that objects of different masses fall at the same speed. A famous story
about Galileo, which was first told in an early biography, recounts that he
performed an experiment from the top of the Leaning Tower of Pisa in
which he threw a cannonball and a smaller wooden ball off the tower at
the same time. His intent, reputedly, was to disprove an assertion attributed to Aristotle that heavier objects would fall faster than light ones.
The account has long been doubted, and it seems pretty clear now that
Galileo never did perform this experiment, but it still makes for a good
story—such a good story that the commander of the Apollo 15 Moon
mission, David Scott, famously dropped a hammer and a falcon feather
onto the surface of the Moon at the same time to see if objects of different mass would fall to the ground at the same rate in a vacuum. It's a
wonderful video, which you can access here: http://video.google.com/
videoplay?docid=6926891572259784994#.

The striking thing to me about this video is just how *slowly* they both
drop. Without thinking about it, you might expect them both to drop
quickly, at least surely the hammer. But they both fall slowly because the
gravitational acceleration on the Moon is about six times less than it is
on Earth.

---

*1 kilogram is about 2.2 pounds.

Why was Galileo right that two objects of different mass would land at the same time? The reason is that the gravitational acceleration is the same for all objects. According to $F = ma$, the larger the mass, the larger the gravitational force, but the acceleration is the same for all objects. Thus they reach the ground with the same speed. Of course, the object with the larger mass will have more energy and will therefore have a greater impact.

Now it's important to note here that the feather and the hammer would not land at the same time if you performed this experiment on Earth. This is the result of air drag, which we've discounted until now. Air drag is a force that opposes the motion of moving objects. Also wind would have much more effect on the feather than on the hammer.

This brings us to a very important feature of the second law. The word *net* in the equation as given above is vital, as nearly always in nature more than one force is acting on an object; all have to be taken into account. This means that the forces have to be added. Now, it's not really as simple as this, because forces are what we call vectors, meaning that they have a magnitude as well as a direction, which means that you cannot really make a calculation like $2 + 3 = 5$ for determining the net force. Suppose only two forces act on a mass of 4 kilograms; one force of 3 newtons is pointing upward, and another of 2 newtons is pointing downward. The sum of these two forces is then 1 newton in the upward direction and, according to Newton's second law, the object will be accelerated upward with an acceleration of 0.25 meters per second per second.

The sum of two forces can even be zero. If I place an object of mass m on my table, according to Newton's second law, the gravitational force on the object is then mg (mass × gravitational acceleration) newtons in the downward direction. Since the object is not being accelerated, the net force on the object must be zero. That means that there must be another force of mg newtons upward. That is the force with which the table pushes upward on the object. A force of mg down and one of mg up add up to a force of zero!

This brings us to Newton's third law: "To every action there is always an equal and opposite reaction." This means that the force that two objects exert on each other are always equal and are directed in opposite directions. As I like to put it, action equals minus reaction, or, as it's known more popularly, "For every action there is an equal and opposite reaction."

Some of the implications of this law are intuitive: a rifle recoils backward against your shoulder when it fires. But consider also that when you push against a wall, it pushes back on you in the opposite direction with the exact same force. The strawberry shortcake you had for your birthday pushed down on the cake plate, which pushed right back at it with an equal amount of force. In fact, odd as the third law is, we are completely surrounded by examples of it in action.

Have you ever turned on the faucet connected to a hose lying on the ground and seen the hose snake all over the place, maybe spraying your little brother if you were lucky? Why does that happen? Because as the water is pushed out of the hose, it also pushes back on the hose, and the result is that the hose is whipped all around. Or surely you've blown up a balloon and then let go of it to see it fly crazily around the room. What's happening is that the balloon is pushing the air out, and the air coming out of the balloon pushes back on the balloon, making it zip around, an airborne version of the snaking garden hose. This is no different from the principle behind jet planes and rockets. They eject gas at a very high speed and that makes them move in the opposite direction.

Now, to truly grasp just how strange and profound an insight this is, consider what Newton's laws tell us is happening if we throw an apple off the top of a thirty-story building. We know the acceleration will be g, about 9.8 meters per second per second. Now, say the apple is about half a kilogram (about 1.1 pounds) in mass. Using the second law, $F = ma$, we find that the Earth attracts the apple with a force of $0.5 \times 9.8 = 4.9$ newtons. So far so good.

But now consider what the third law demands: if the Earth attracts the apple with a force of 4.9 newtons, then the apple will attract the Earth with a force of 4.9 newtons. Thus, as the apple falls to Earth, the Earth falls to the apple. This seems ridiculous, right? But hold on. Since the mass of the Earth is so much greater than that of the apple, the numbers get pretty wild. Since we know that the mass of the Earth is about 6×10^{24} kilograms, we can calculate how far it falls up toward the apple: about 10^{-22} meters, about one ten-millionth of the size of a proton, a distance so small it cannot even be measured; in fact, it's meaningless.

This whole idea, that the force between two bodies is both equal and in opposite directions, is at play everywhere in our lives, and it's the key to why your scale goes berserk when you lift yourself up onto your toes on it. This brings us back to the issue of just what weight is, and lets us understand it more precisely.

When you stand on a bathroom scale, gravity is pulling down on you with force mg (where m is your mass) and the scale is pushing up on you with the same force so that the net force on you is zero. This force pushing up against you is what the scale actually measures, and this is what registers as your weight. Remember, weight is not the same thing as mass. For your mass to change, you'd have to go on a diet (or, of course, you might do the opposite, and eat more), but your weight can change much more readily.

Let's say that your mass (m) is 55 kilograms (that's about 120 pounds). When you stand on a scale in your bathroom, you push down on the scale with a force mg, and the scale will push back on you with the same force, mg. The net force on you is zero. The force with which the scale pushes back on you is what you will read on the scale. Since your scale may indicate your weight in pounds, it will read 120 pounds.

Let's now weigh you in an elevator. While the elevator stands still (or while the elevator is moving at constant speed), you are not being accelerated (neither is the elevator) and the scale will indicate that you weigh 120 pounds, as was the case when you weighed yourself in your bathroom. We enter the elevator (the elevator is at rest), you go on the scale, and it reads 120 pounds. Now I press the button for the top floor, and the elevator briefly accelerates upward to get up to speed. Let's assume that this acceleration is 2 meters per second per second and that it is constant. During the brief time that the elevator accelerates, the net force on you cannot be zero. According to Newton's second, the net force F_{net} on you must be $F_{net} = ma_{net}$. Since the net acceleration is 2 meters per second per second, the net force on you is $m \times 2$ upward. Since the force of gravity on you is mg down, there must be a force of $mg + m2$, which can also be written as $m(g + 2)$, on you in upward direction. Where does this force come from? It must come from the scale (where else?). The scale is exerting a force $m(g + 2)$ on you upward. But remember that the weight that the scale indicates is the force with which it pushes upward on you. Thus the scale tells you that your weight is about 144 pounds (remember, g is about 10 meters per second per second). You have gained quite a bit of weight!

According to Newton's third, if the scale exerts a force of $m(g + 2)$ on you upward, then you must exert the same force on the scale downward. You may now reason that if the scale pushes on you with the same force that you push on the scale, that then the net force on you is zero, thus you cannot be accelerated. If you reason this way, you make a very common mistake. There are only two forces acting on you: mg down due to gravity and $m(g + 2)$ up due to the scale, and thus a net force of $2m$ is exerted on you in an upward direction, which will accelerate you at 2 meters per second per second.

The moment the elevator stops accelerating, your weight goes back to normal. Thus it's only during the short time of the upward acceleration that your weight goes up.

You should now be able to figure out on your own that if the elevator is being accelerated downward, you lose weight. During the time that the acceleration downward is 2 meters per second per second, the scale will register that your weight is $m(g - 2)$, which is about 96 pounds. Since an elevator that goes up must come to a halt, it must be briefly accelerated downward before it comes to a stop. Thus near the end of your elevator ride up you will see that you lost weight, which you may enjoy! However, shortly after that, when the elevator has come to a stop, your weight will again go back to normal (120 pounds).

Suppose now, someone who really, really dislikes you cuts the cable and you start zooming down the elevator shaft, going down with an acceleration of g. I realize you probably wouldn't be thinking about physics at that point, but it would make for a (briefly) interesting experience. Your weight will become $m(g - g) = 0$; you are weightless. Because the scale is falling downward at the same acceleration as you, it no longer exerts a force on you upward. If you looked down at the scale it would register zero. In truth, you would be floating, and everything in the elevator would be floating. If you had a glass of water you could turn it over and the water would not fall out, though of course this is one experiment I urge you not to try!

This explains why astronauts float in spaceships. When a space module, or the space shuttle, is in orbit, it is actually in a state of free fall, just like the free fall of the elevator. What exactly is free fall? The answer might surprise you. Free fall is when the force acting upon you is exclusively gravitational, and no other forces act on you. In orbit, the astronauts, the spaceship, and everything inside it are all falling toward Earth in free fall. The reason why the astronauts don't go splat is because the Earth is curved and the astronauts, the spaceship, and everything inside it are moving so fast that as they fall toward Earth, the surface of the planet curves away from them, and they will never hit the Earth's surface.

Thus the astronauts in the shuttle are weightless. If you were in the shuttle, you would think that there is no gravity; after all, nothing in the shuttle has any weight. It's often said that the shuttle in orbit is a zero-gravity environment, since that's the way you perceive it. However, if there were no gravity, the shuttle would not stay in orbit.

The whole idea of changing weight is so fascinating that I really wanted to be able to demonstrate this phenomenon—even weightlessness—in class. What if I climbed up on a table, standing on a bathroom scale that was tied very securely to my feet? I thought then maybe I could somehow show my students—by rigging up a special camera—that for the half second or so that I was in free fall the bathroom scale would indicate zero. I might recommend that you try this yourself, but don't bother; trust me, I tried it many times and only broke many scales. The problem is that the scales you can buy commercially don't react nearly fast enough, since there is inertia in their springs. One of Newton's laws bedeviling another! If you could jump off a thirty-story building, you would probably have enough time (you would have about 4.5 seconds) to see the effect, but of course there would be other problems with that experiment.

So rather than breaking scales or jumping off buildings, here's something you can try in your backyard to experience weightlessness, if you have a picnic table and good knees. I do this from the lab table in front of my classroom. Climb up on the table and hold a gallon or half-gallon jug of water in your outstretched hands, just cradling it lightly on top of them, not holding the sides of the jug. It has to be just resting on your hands. Now jump off the table, and while you are in the air you will see the jug start floating above your hands. If you can get a friend to make a digital video of you taking the jump, and play it back in slow motion, you will very clearly see the jug of water start to float. Why? Because as you accelerate downward the force with which you have been pushing up on the jug, to keep it in your hands, has become zero. The jug will now be accelerated at 9.8 meters per second per second, just as you are. You and the jug are both in free fall.

But how does all of this explain why your scale goes berserk when you lift yourself up on your toes? As you push yourself upward you accelerate upward, and the force of the scale pushing on you increases. So you weigh more for that brief time. But then, at the top of your toes, you decelerate to come to a halt, and that means that your weight goes down. Then, when you let your heels down, the entire process is reversed, and you have just demonstrated how, without changing your mass at all, you can make yourself weigh more or less for fractions of a second.

The Law of Universal Gravitation:
Newton and the Apple

People commonly refer to Newton's three laws, but, in fact, he formulated four. We've all heard the story of Newton observing an apple falling from a tree one day in his orchard. One of Newton's early biographers claimed that Newton himself told the story. "It was occasion'd by the fall of an apple," wrote Newton's friend William Stukeley, quoting a conversation he had with Newton, "as he sat in contemplative mood. Why should that apple always descend perpendicularly to the ground, thought he to himself." * But many remain unconvinced that the story is true. After all, Newton only told Stukeley the story a year before he died, and he made no mention of it any other place in his voluminous writings.

Still, what is unquestionably true is that Newton was the first to realize that the same force that causes an apple to fall from a tree governs the motion of the Moon, the Earth, and the Sun—indeed, of all the objects in the universe. That was an extraordinary insight, but once again, he didn't stop there. He realized that every object in the universe attracts every other object—and he came up with a formula for calculating just how strong the attraction is, known as his universal law of gravitation. This law states that the force of gravitational attraction between two objects is directly proportional to the product of the masses of the objects and inversely proportional to the square of the distance between them.

So, in other words, to use a purely hypothetical example, which I stress has no relation to reality, if Earth and Jupiter were orbiting the Sun at the same distance, then because Jupiter is about 318 times more massive than Earth the gravitational force between the Sun and Jupiter would be about 318 times greater than that between the Sun and Earth. And if Jupiter and Earth were the same mass, but Jupiter were

---

*The Royal Society recently posted a digital image of Stukeley's manuscript online, which you can find here: http://royalsociety.org/turning-the-pages/.

in its actual orbit, which is about five times farther from the Sun than the Earth's orbit, then because the gravitational force is inversely proportional to the square of the distance, it would be twenty-five times greater between the Sun and Earth than between the Sun and Jupiter.

In Newton's famous *Philosophiæ Naturalis Principia Mathematica* published in 1687—which we now call the *Principia*—he did not use an equation to introduce the law of universal gravitation, but that's the way we express it most often in physics today:

$$F_{grav} = G \frac{m_1 m_2}{r^2}$$

Here, F_{grav} is the force of gravitational attraction between an object of mass m_1 and one of mass m_2, and r is the distance between them; the 2 means "squared." What is G? That's what's called the gravitational constant. Newton knew, of course, that such a constant exists, but it is not mentioned in his *Principia*. From the many measurements that have since been done, we now know that the most accurate value for G is $6.67428 \pm 0.00067 \times 10^{-11}$.* We physicists also do believe that it's the same throughout the universe, as Newton conjectured.

The impact of Newton's laws was gigantic and cannot be overestimated; his *Principia* is among the most important works of science ever written. His laws changed all of physics and astronomy. His laws made it possible to calculate the mass of the Sun and planets. The way it's done is immensely beautiful. If you know the orbital period of any planet (say, Jupiter or the Earth) and you know its distance to the Sun, you can calculate the mass of the Sun. Doesn't this sound like magic? We can carry this one step further; if you know the orbital period of one of Jupiter's bright moons (discovered by Galileo in 1609) and you know the distance between Jupiter and that moon, you can calculate the mass of Jupiter. Therefore, if you know the orbital period of the Moon around the Earth (it's 27.32 days) and you know the mean distance between the Earth

---

*If you ever want to use this value, make sure that your masses are in kilograms and that the distance, r, is in meters. The gravitational force will then be in newtons.

and the Moon (it's about 239,000 miles) then you can calculate to a high degree of accuracy the mass of the Earth. I show you how this works in appendix 2. If you can handle some math you may enjoy it!

But Newton's laws reach far beyond our solar system. They dictate and explain the motion of stars, binary stars (chapter 13), star clusters, galaxies, and even clusters of galaxies, and Newton's laws deserve credit for the twentieth-century discovery of what we call dark matter. I will tell you more about this later. His laws are beautiful—breathtakingly simple and incredibly powerful at the same time. They explain so much, and the range of phenomena they clarify is mind-boggling.

By bringing together the physics of motion, of interactions between objects, and of planetary movements, Newton brought a new kind of order to astronomical measurements, showing how what had been a jumble of confusing observations made through the centuries were all interconnected. Others had had glimmers of his insights, but they hadn't been able to put them together as he did.

Galileo, who died the year before Newton was born, had come up with an early version of Newton's first law and could describe the motion of many objects mathematically. He also discovered that all objects will fall from a given height at the same speed (in the absence of air drag). He couldn't, though, explain *why* it was true. Johannes Kepler had worked out the fundamentals of *how* planetary orbits worked, but he had no clue *why*. Newton explained the why. And, as we've seen, the answers, and many of the conclusions they lead to, are not in the slightest bit intuitive.

The forces of motion are endlessly fascinating to me. Gravity is always with us; it pervades the universe. And the astounding thing about it— well, one astounding thing—is that it acts at a distance. Have you ever really stopped to consider that our planet stays in orbit, that we are all alive because of the attractive force between two objects 93 million miles apart?

Pendulums in Motion

Even though gravity is a pervasive force in our lives, there are many ways in which the effects it has on our world confound us. I use a pendulum demonstration to surprise students with just how counterintuitively gravity operates. Here's how it works.

Many of you may think that if you swing on a playground swing next to someone who is much lighter than you are, e.g., a toddler, you'll go much slower than that person. But that is not the case. It may therefore come as a surprise to you that the amount of time it takes to complete one swing of a pendulum, which we call the period of the pendulum, is not affected by the weight hanging from the pendulum (we call this weight the bob). Note that here I'm talking about what's called a simple pendulum, which means that it meets two conditions. First, the weight of the bob must be so much larger than the weight of the string that the weight of the string can be ignored. Second, the size of the bob needs to be small enough that we can treat it as if it were just a point, which has zero size.* It's easy to make a simple pendulum at home: attach an apple to the end of a lightweight string that is at least four times longer than the size of the apple.

Using Newton's laws of motion, I derive in class an equation for calculating the period of a simple pendulum, and then I put the equation to the test. To do that I have to make the assumption that the angle over which the pendulum swings is small. Let me be more precise about what I mean by that. When you look at your homemade pendulum as it swings back and forth, from right to left and from left to right, you will see that most of the time the pendulum is moving, either to the left or to the right. However, there are two times during a complete swing that the pendulum stands still, after which it reverses direction. When

---

*If the mass of the string cannot be ignored, and/or if the size of the bob cannot be treated as a point mass, then it is no longer a simple pendulum. We call it a physical pendulum and it behaves differently.

this happens the angle between the string and the vertical has reached a maximum value, which we call the amplitude of the pendulum. If air drag (friction) can be ignored, that maximum angle when the pendulum comes to a halt at the far left is the same as when the pendulum comes to a halt at the far right. The equation that I derive is only valid for small angles (small amplitudes). We call such a derivation in physics a small-angle approximation. Students always ask me, "How small is small?" One student is even very specific; she asks, "Is an amplitude of five degrees small? Is the equation still valid for an amplitude of ten degrees or is ten degrees not small?" Of course, those are excellent questions, and I suggest that we will bring this to a test in class.

The equation that I derive is quite simple and very elegant, though it may look a little daunting to those who haven't been doing any math lately: $T = 2\pi \sqrt{\dfrac{L}{g}}$

T is the period of the pendulum (in seconds), L is the length of the string (in meters), π is 3.14, and g is the gravitational acceleration (9.8 meters per second per second). So the right part of the equation reads two π multiplied by the square root of the length of the string divided by the gravitational acceleration. I won't go into the details here of why this is the correct equation (you can follow the derivation that I do in my recorded lectures if you want to; the website link is on page 54).

I am giving the equation here so that you can appreciate just how precisely my demonstrations confirm it. The equation predicts that a pendulum 1 meter long has a period of about 2 seconds. I measure the time it takes a pendulum, with a string that long, to complete ten oscillations, and that comes to about 20 seconds. Dividing by 10, we get 2 seconds for the period. Then I go to a pendulum with a string that is four times shorter. The equation predicts that the period should be twice as short. So I make the string 25 centimeters long, and indeed it takes about 10 seconds for ten oscillations. So that is all very reassuring.

To bring the equation to a much more careful test than what I did with the handheld small apple pendulum, I had a simple pendulum constructed in my classroom: a rope 5.18 meters (about 17 feet) long with a spherical steel bob weighing 15 kilograms at the end of the rope. I call it the mother of all pendulums. You can see it near the end of my lecture here: http://ocw.mit.edu/courses/physics/8-01-physics-i-classical -mechanics-fall-1999/video-lectures/embed10/.

What should the period, T, of this pendulum be? $T = 2\pi\sqrt{\frac{5.18}{9.8}}$, which is 4.57 seconds. To bring this to a test, as I promised my students, I measure the period both for a 5-degree and for a 10-degree amplitude.

I use a large digital timer that the students can see, and that displays the time to an accuracy of one-hundredth of a second. I've tested my reaction time in turning the timer on and off countless times over the years, and I know it's about one-tenth of a second (on a good day). This means that if I repeat the very same measurement a dozen times I will get measurements for the period that will vary by as much as 0.1 (maybe 0.15) seconds. So whether I measure the time it takes for one oscillation or for ten oscillations, my timing will have an uncertainty of plus or minus 0.1 seconds. I therefore let the pendulum swing ten times, as that will give a ten times more accurate value for the period than if I let it swing only once.

I pull the bob out enough so that the angle of the rope with the vertical is about 5 degrees and then let it go and start the timer. The class counts each of the swings out loud, and after ten oscillations I stop the timer. It's amazing—the timer reads 45.70 seconds, ten times my estimate for one swing. The class applauds wildly.

Then I increase the amplitude to 10 degrees, let the bob go, start the timer, get the class counting, and right at ten, I stop the timer: 45.75 seconds. 45.75 ± 0.1 seconds for ten oscillations translates into 4.575 ± 0.01 seconds per oscillation. The result for the 5-degree amplitude is the same as for the 10-degree amplitude (within the uncertainty of the measurements). So my equation is still very accurate.

Then I ask the class, Suppose I sat on the bob and swung along with it—would we get the same period, or would it change? I never look forward to sitting on this thing (it really hurts), but for science, and to get the students laughing and involved, I wouldn't miss the opportunity. Of course I can't sit upright on the bob because that way I will effectively shorten the rope, and reduce the period a bit. But if I make my body as horizontal as possible in order to be at the same level as the bob, I keep the rope length pretty much the same. So I pull the bob up, put it between my legs, grasp the rope, and let myself go. You can see this on the jacket of this book!

It's not easy for me to start and stop the timer while hanging on the pendulum without increasing my reaction time. However, I've practiced this so many times that I am quite sure that I can achieve an uncertainty in my measurements of ± 0.1 seconds. I swing ten times, with students counting the swings out loud—and laughing at the absurdity of my situation while I complain and groan loudly—and when after ten oscillations I turn off the timer, it reads 45.61 seconds. That's a period of 4.56 ± 0.01 seconds. "Physics works!" I scream, and the students go bananas.

Grandmothers and Astronauts

Another tricky aspect of gravity is that we can be fooled into perceiving that it's pulling from a different direction than it really is. Gravity always pulls toward the center of Earth—on Earth, that is, not on Pluto of course. But we can sometimes perceive that gravity is operating horizontally, and this artificial or perceived gravity, as we call it, can in fact seem to defy gravity itself.

You can demonstrate this artificial gravity easily by doing something my grandmother used to do every time she made a salad. My grandmother had such fantastic ideas—remember, she's the one who taught me that you're longer when you're lying down than when you're standing up. Well, when she made a salad, she really had a good time. She would wash the lettuce in a colander, and then rather than drying it in a cloth towel, which would damage the leaves, she had invented her own technique: she took the colander and put a dish towel over the top, holding it in place with a rubber band, and then she would swing it around furiously in a circle—I mean really fast.

That's why when I demonstrate this in class, I make sure to tell the students in the first two rows to close their notebooks so their pages don't get wet. I bring lettuce into the classroom, wash it carefully in the sink on my table, prepare it in the colander. "Get ready," I tell them, and I swing my arm vigorously in a vertical circle. Water drops spray everywhere! Now, of course, we have boring plastic salad spinners to substitute for my grandmother's method—a real pity in my book. So much of modern life seems to take the romance out of things.

This same artificial gravity is experienced by astronauts as they accelerate into orbit around the Earth. A friend and MIT colleague of mine, Jeffrey Hoffman, has flown five missions in the space shuttle, and he tells me that the crew experiences a range of different accelerations in the course of a launch, from about $0.5g$ initially, building to about $2.5g$ at the end of the solid fuel stage. Then it drops back down to about $1g$ briefly, at which point the liquid fuel starts burning, and acceleration builds back up to $3g$ for the last minute of the launch—which takes about eight and a half minutes total to obtain a speed of about 17,000 miles per hour. And it's not at all comfortable. When they finally reach orbit they become weightless and they perceive this as zero gravity.

As you now know, both the lettuce, feeling the colander pushing against it, and the astronauts, feeling the seats pushing against them, are experiencing a kind of artificial gravity. My grandmother's contraption— and our salad spinners—are of course versions of a centrifuge, separating the lettuce from the water clinging to its leaves, which shoots out through the colander's holes. You don't have to be an astronaut to experience this perceived gravity. Think of the fiendish ride at amusement parks called the Rotor, in which you stand at the edge of a large rotating turntable with your back against a metal fence. As it starts to rotate faster and faster, you feel more and more pushed into the fence, right? According to Newton's third law, you push on the wall with the same force as the wall pushes on you.

This force with which the wall pushes on you is called the centripetal force. It provides the necessary acceleration for you to go around; the faster you go, the larger is the centripetal force. Remember, if you go around in a circle, a force (and therefore an acceleration) is required even if the speed remains unchanged. In similar fashion, gravity provides the centripetal force on planets to go around the Sun, as I discuss in appendix 2. The force with which you push on the wall is often called the centrifugal force. The centripetal force and the centrifugal force have the same magnitude but in opposite direction. Do not confuse the two. It's *only* the centripetal force that acts on you (not the centrifugal force), and it is only the centrifugal force that acts on the wall (not the centripetal force).

Some Rotors can go so fast that they can lower the floor on which you stand and you won't slide down. Why won't you slide down?

Think about it. If the Rotor isn't spinning at all the force of gravity on you will make you slide down as the frictional force between you and the wall (which will be upward) is not large enough to balance the force of gravity. However, the frictional force, with the floor lowered, will be higher when the Rotor spins, as it depends on the centripetal force. The larger the centripetal force (with the floor lowered), the larger the frictional force. Thus, if the Rotor spins fast enough with the floor lowered, the frictional force can be large enough that it will balance the force of gravity and thus you won't slide down.

There are lots of ways to demonstrate artificial gravity. Here's one you can try at home; well, in your backyard. Tie a rope to the handle of an empty paint can and fill the can with water—about half full, I'd say, otherwise it will be awfully heavy to spin—and then whip the can around as hard as you can up over your head in a circle. It might take some practice to get it going fast enough. Once you do, you'll see that not a drop of water will fall out. I have students do this in my classes, and I must say it's a complete riot! This little experiment also explains why, with some especially pernicious versions of the Rotor, the floor will gradually turn over until you are completely upside down at one point, and yet you don't drop down to the ground (of course, for safety's sake, you are also strapped into the thing).

The force with which a scale pushes on us determines what the scale tells us we weigh; it's the force of gravity—not the lack of it—that makes astronauts weightless; and when an apple falls to Earth, the Earth falls to the apple. Newton's laws are simple, far-reaching, profound, and utterly counterintuitive. In working out his famous laws, Sir Isaac Newton was contending with a truly mysterious universe, and we have all benefited enormously from his ability to unlock some of these mysteries and to make us see our world in a fundamentally new way.

CHAPTER 4

The Magic of Drinking with a Straw

One of my favorite in-class demonstrations involves two paint cans and a rifle. I fill one can to the rim with water and then bang the top on tightly. Then I fill the second can most of the way, but leaving an inch or so of space below the rim, and also seal that one. After placing them one in front of the other on a table, I walk over to a second table several yards away, on which rests a long white wooden box, clearly covering some kind of contraption. I lift up the box, revealing a rifle fastened onto a stand, pointing at the paint cans. The students' eyes widen—am I going to fire a rifle in class?

"If we were to shoot a bullet through these paint cans, what would happen?" I ask them. I don't wait for answers. I bend down to check the rifle's aim, usually fiddling with the bolt a little. This is good for building up tension. I blow some dust out of the chamber, slide a bullet in, and announce, "All right, there goes the bullet. Are we ready for this?" Then standing alongside the rifle, I put my finger on the trigger, count "Three, two, one"—and fire. One paint can's top instantly pops way up into the air, while the other one stays put. Which can do you think loses its top?

To know the answer, you first have to know that air is compressible and water isn't; air molecules can be squished closer in toward one another, as can the molecules of any gas, but those of water—and of any liquid at all—cannot. It takes horrendous forces and pressures to change the density of a liquid. Now, when the bullet enters the paint cans, it brings a great deal of pressure with it. In the can with the air in it, the air acts like a cushion, or a shock absorber, so the water isn't disturbed and the can doesn't explode. But in the can full of water, the water can't compress. So the extra pressure the bullet introduces in the water exerts a good deal of force on the walls and on the top of the can and the top blows off. As you may imagine, it's really very dramatic and my students are always quite shocked.

Surrounded by Air Pressure

I always have a lot of fun with pressure in my classes, and air pressure is particularly entertaining because so much is so counterintuitive about it. We don't even realize we are experiencing air pressure until we actually look for it, and then it's just astonishing. Once we realize it's there—and begin to understand it—we begin to see evidence for it everywhere, from balloons to barometers, to why a drinking straw works, to how deep you can swim and snorkel in the ocean.

The things we don't see at first, and take for granted, like gravity and air pressure, turn out to be among the most fascinating of all phenomena. It's like the joke about two fish swimming along happily in a river. One fish turns to the other, a skeptical look on its face, and says, "What's all this new talk about 'water'?"

In our case, we take the weight and density of our invisible atmosphere for granted. We live, in truth, at the bottom of a vast ocean of air, which exerts a great deal of pressure on us every second of every day. Suppose I hold my hand out in front of me, palm up. Now imagine a very long piece of square tubing that is 1 centimeter wide (on each side, of course) balanced on my hand and rising all the way to the top of the atmosphere. That's more than a hundred miles. The weight of the air alone in the tube—forget about the tubing—would be about 1 kilogram, or about 2.2 pounds.* That's one way to measure air pressure: 1.03 kilograms per square centimeter of pressure is called the standard atmosphere. (You may also know it as about 14.7 pounds per square inch.)

Another way to calculate air pressure—and any other kind of pressure—is with a fairly simple equation, one so simple that I've actually just put it in words without saying it was an equation. Pressure is force divided by area: $P = F/A$. So, air pressure at sea level is about 1 kilogram per square centimeter. Here's another way to visualize the relationship between force, pressure, and area.

Suppose you are ice-skating on a pond and someone falls through. How do you approach the hole—by walking on the ice? No, you get down on your stomach and slowly inch forward, distributing the force of your body on the ice over a larger area, so that you put less pressure on the ice, making it much less likely to break. The difference in pressure on the ice when standing versus lying down is remarkable.

Say you weigh 70 kilograms and are standing on ice with two feet planted. If your two feet have a surface area of about 500 square centimeters (0.05 square meters), you are exerting 70/0.05 kilograms per square meter of pressure, or 1,400 kilograms per square meter. If you lift up one foot, you will have doubled the pressure to 2,800 kilograms per square meter. If you are about 6 feet tall, as I am, and lie down on the ice, what happens? Well, you spread the 70 kilograms over about 8,000 square centimeters, or about 0.8 square meters, and your body exerts just 87.5 kilograms per square meter of pressure, roughly thirty-two times less than while you were standing on one foot. The larger the area, the lower the pressure, and, conversely, the smaller the area, the larger the pressure. Much about pressure is counterintuitive.

For example, pressure has no direction. However, the force caused by pressure does have a direction; it's perpendicular to the surface the pres-

---

*Remember, all you scientists, I'm using common rather than technical language here. Even though a kilogram is in fact a unit of mass, not weight, it's often used for both, and that's what I'm doing here.

sure is acting on. Now stretch out your hand (palm up) and think about the force exerted on your hand—no more tube involved. The area of my hand is about 150 square centimeters, so there must be a 150-kilogram force, about 330 pounds, pushing down on it. Then why am I able to hold it up so easily? After all, I'm no weight lifter. Indeed, if this were the only force, you would not be able to carry that weight on your hand. But there is more. Because the pressure exerted by air surrounds us on all sides, there is also a force of 330 pounds upward on the back of your hand. Thus the net force on your hand is zero.

But why doesn't your hand get crushed if so much force is pressing in on it? Clearly the bones in your hand are more than strong enough not to get crushed. Take a piece of wood of the size of your hand; it's certainly not getting crushed by the atmospheric pressure.

But how about my chest? It has an area of about 1,000 square centimeters. Thus the net force exerted on it due to air pressure is about 1,000 kilograms: 1 metric ton. The net force on my back would also be about 1 ton. Why don't my lungs collapse? The reason is that inside my lungs the air pressure is also 1 atmosphere; thus, there is no pressure difference between the air inside my lungs and the outside air pushing down on my chest. That's why I can breathe easily. Take a cardboard or wooden or metal box of similar dimensions as your chest. Close the box. The air inside the box is the air you breathe—1 atmosphere. The box does not get crushed for the same reason that your lungs will not collapse. Houses do not collapse under atmospheric pressure because the air pressure inside is the same as outside; we call this pressure equilibrium. The situation would be very different if the air pressure inside a box (or a house) were much lower than 1 atmosphere; chances are it would then get crushed, as I demonstrate in class. More about this later.

The fact that we don't normally notice air pressure doesn't mean it's not important to us. After all, weather forecasts are constantly referring to low- and high-pressure systems. And we all know that a high-pressure system will tend to bring nice clear days, and a low-pressure system means some kind of storm front is approaching. So measuring air pressure is something we very much want to do—but if we can't feel it, how do we do that? You may know that we do it with a barometer, but of course that doesn't explain much.

The Magic of Straws

Let's begin with a little trick that you've probably done dozens of times. If you put a straw into a glass of water—or as I like to do in class, of cranberry juice—it fills up with juice. Then, if you put a finger over the top of the straw and start pulling it out of the glass, the juice stays in the straw; it's almost like magic. Why is this? The explanation is not so simple.

In order to explain how this works, which will help us get to a barometer, we need to understand pressure in liquids. The pressure caused by liquid alone is called hydrostatic pressure ("hydrostatic" is derived from the Latin for "liquid at rest"). Note that the total pressure below the surface of a liquid—say, the ocean—is the total of the atmospheric pressure above the water's surface (as with your outstretched hand) and the hydrostatic pressure. Now here's a basic principle: *In a given liquid that is stationary, the pressure is the same at the same levels. Thus the pressure is everywhere the same in horizontal planes.*

So if you are in a swimming pool, and you put your hand 1 meter below the surface of the pool at the shallow end, the total pressure on your hand, which is the sum of the atmospheric pressure (1 atmosphere) and the hydrostatic pressure, will be identical to the pressure on your friend's hand, also at 1 meter below the surface, at the deep end of the pool. But if you bring your hand down to 2 meters below the surface, it will experience a hydrostatic pressure that is twice as high. The more fluid there is above a given level, the greater the hydrostatic pressure at that level.

The same principle holds true for air pressure, by the way. Sometimes we talk about our atmosphere as being like an ocean of air, and at the bottom of this ocean, over most of Earth's surface, the pressure is about 1 atmosphere. But if we were on top of a very tall mountain, there would be less air above us, so the atmospheric pressure would be less. At the summit of Mount Everest, the atmospheric pressure is only about one third of an atmosphere.

Now, if for some reason the pressure is not the same in a horizontal plane, then the liquid will flow until the pressure in the horizontal plane is equalized. Again, it's the same with air, and we know the effect as wind—it's caused by air moving from high pressure to low pressure to even out the differences, and it stops when the pressure is equalized.

So what's happening with the straw? When you lower a straw into liquid—for now with the straw open at the top—the liquid enters the straw until its surface reaches the same level as the surface of the liquid in the glass outside the straw; the pressure on both surfaces is the same: 1 atmosphere.

Now suppose I suck on the straw. I will take some of the air out of it, which lowers the pressure of the column of air above the liquid inside the straw. If the liquid inside the straw remained where it was, then the pressure at its surface would become lower than 1 atmosphere, because the air pressure above the liquid has decreased. Thus the pressure on the two surfaces, inside and outside the straw, which are *at the same level* (in the same horizontal plane) would differ, and that is not allowed. Consequently, the liquid in the straw rises until the pressure in the liquid inside the straw at the same level as the surface outside the straw again becomes 1 atmosphere. If by sucking, I lower the air pressure in the straw by 1 percent (thus from 1.00 atmosphere to 0.99 atmosphere) then just about any liquid we can think of drinking—water or cranberry juice or lemonade or beer or wine—would rise about 10 centimeters. How do I know?

Well, the liquid in the straw has to rise to make up for the 0.01-atmosphere loss of air pressure above the liquid in the straw. And from the formula for calculating the hydrostatic pressure in a liquid, which I won't go into here, I know that a hydrostatic pressure of 0.01 atmosphere for water (or for any comparably dense liquid) is created by a column of 10 centimeters.

If the length of your straw was 20 centimeters, you would have to suck hard enough to lower the air pressure to 0.98 atmosphere in order for the juice to rise 20 centimeters and reach your mouth. Keep this in mind for later. Now that you know all about weightlessness in the space shuttle (chapter 3) and about how straws work (this chapter), I have an interesting problem for you: A ball of juice is floating in the shuttle. A glass is not needed as the juice is weightless. An astronaut carefully inserts a straw into the ball of juice, and he starts sucking on the straw. Will he be able to drink the juice this way? You may assume that the air pressure in the shuttle is about 1 atmosphere.

Now back to the case of the straw with your finger on top. If you raise the straw slowly up, say 5 centimeters, or about 2 inches, as long as the straw is still in the juice, the juice will not run out of the straw. In fact it will almost (not quite) stay exactly at the mark where it was before. You can test this by marking the side of the straw at the juice line before you lift it. The surface of the juice inside the straw will now be about 5 centimeters higher than the surface of the juice in the glass.

But given our earlier sacred statement about the pressure equalizing inside and outside of the straw—at the same level—how can this be? Doesn't this violate the rule? No it does not! Nature is very clever; the air trapped by your finger in the straw will increase its volume just enough so that its pressure will decrease just the right amount (about 0.005 atmosphere) so that the pressure in the liquid in the straw at the same level of the surface of the liquid in the glass becomes the same: 1 atmosphere. This is why the juice will not rise precisely 5 centimeters, but rather just a little less, maybe only 1 millimeter less—just enough to give the air enough extra volume to lower its pressure to the desired amount.

Can you guess how high water (at sea level) can go in a tube when you've closed off one end and you slowly raise the tube upward? It depends on how much air was trapped inside the tube when you started raising it. If there was very little air in the straw, or even better no air at all, the maximum height the water could go would be about 34 feet—a little more than 10 meters. Of course, you couldn't do this with a small glass, but a bucket of water might do. Does this surprise you? What makes it even more difficult to grasp is that the shape of the tube doesn't matter. You could make it twist and even turn it into a spiral, and the water can still reach a vertical height of 34 feet, because 34 feet of water produces a hydrostatic pressure of 1 atmosphere.

Knowing that the lower the atmospheric pressure, the lower the maximum possible column of water will be, provides us with a way to measure atmospheric pressure. To see this, we could drive to the top of Mount Washington (about 6,300 feet high), where the atmospheric pressure is about 0.82 atmosphere, so this means that the pressure at the surface outside the tube is no longer 1 atmosphere but only about 0.82 atmosphere. So, when I measure the pressure in the water inside the tube at the level of the water surface outside the tube, it must also be 0.82 atmosphere, and thus the maximum possible height of the water column will be lower. The maximum height of water in the tube would then be 0.82 times 34 feet, which is about 28 feet.

If we measure the height of that column using cranberry juice by marking meters and centimeters on the tube, we have created a cranberry juice barometer—which will indicate changes in air pressure. The French scientist Blaise Pascal, by the way, is said to have made a barometer using red wine, which is perhaps to be expected of a Frenchman. The man credited with inventing the barometer in the mid-seventeenth century, the Italian Evangelista Torricelli, who was briefly an assistant to Galileo, settled eventually on mercury for his barometer. This is because, for a given column, denser liquids produce more hydrostatic pressure and so they have to rise less in the tube. About 13.6 times denser than water, mercury made the length of the tube much more convenient. The hydrostatic pressure of a 34-foot column of water (which is 1 atmosphere) is the same as 34 feet divided by 13.6 which is 2.5 feet of mercury (2.5 feet is 30 inches or 76 centimeters).

Torricelli wasn't actually trying to measure air pressure at first with his device. He was trying to find out whether there was a limit to how high suction pumps could draw up a column of water—a serious problem in irrigation. He poured mercury to the top of a glass tube about 1 meter long, closed at the bottom. He then sealed the opening at the rim with his thumb and turned it upside down, into a bowl of mercury, taking his thumb away. When he did this, some of the mercury ran out of the tube back into the bowl, but the remaining column was about 76 centimeters high. The empty space at the top of the tube, he argued, was a vacuum, one of the very first vacuums produced in a laboratory. He knew that mercury was about 13.6 times denser than water, so he could calculate that the maximum length of a water column—which was what he really wanted to know—would be about 34 feet. While he was working this out, as a side benefit, he noticed that the level of the liquid rose and fell over time, and he came to believe that these changes were due to changes in atmospheric pressure. Quite brilliant. And his experiment explains why mercury barometers always have a little extra vacuum space at the top of their tubes.

Pressure Under Water

By figuring out the maximum height of a column of water, Torricelli also figured out something you may have thought about while trying to catch a glimpse of fish in the ocean. My hunch is you've probably tried snorkeling at some point in your life. Well, most snorkels have tubes no more than a foot long; I'm sure you've wanted to go deeper at times and wished the snorkel were longer. How deep do you think you could go and still have the snorkel work? Five feet, ten feet, twenty?

I like to find the answer to this question in class with a simple device called a manometer; it's a common piece of lab equipment. It's very simple, and you could easily make one at home, as I'll describe in just a bit. What I really want to find out is how deep I can be below the surface and still suck air into my lungs. In order to figure this out, we have to measure the hydrostatic pressure of the water bearing in on my chest, which gets more powerful the deeper I go.

The pressure surrounding us, which is, remember, identical at identical levels, is the sum of the atmospheric pressure and the hydrostatic pressure. If I snorkel below the surface of the water, I breathe in air from the outside. That air has a pressure of 1 atmosphere. As a result, when I take air in through the snorkel, the pressure of the air in my lungs becomes the same, 1 atmosphere. But the pressure on my chest is the atmospheric pressure *plus* the hydrostatic pressure. So now the pressure on my chest is *higher* than the pressure inside my lungs; the difference is exactly the hydrostatic pressure. This causes no problem in exhaling, but when I inhale, I have to expand my chest. And if the hydrostatic pressure is too high because I'm too deep in the water, I simply don't have the muscular strength to overcome the pressure difference, and I can't take in more air. That's why, if I want to go deeper in the water, I need to breathe in pressurized air to overcome the hydrostatic pressure. But highly pressurized air is quite taxing on our bodies, which is why there are strict limits to the amount of time for dives.

Now to come back to snorkeling, how far down can I go? To figure this out, I rig a manometer up on the wall of the lecture hall. Imagine a transparent plastic tube about 4 meters long. I attach one end to the wall high up on the left and then snake it into a U shape on the wall. Each arm of the U is a little less than 2 meters long. I pour about 2 meters' worth of cranberry juice into the tube and it naturally settles to the same level on each side of the U tube. Now, by blowing into the right end of the tube I push the cranberry juice up on the left side of the U tube. The vertical distance I can push the juice up will tell me how deep I will be able to snorkel. Why? Because this is a measure of how much pressure my lungs can apply to overcome the hydrostatic pressure of the water—cranberry juice and water being for this purpose equivalent—but the cranberry juice is easier to see for the students.

I lean over, exhale completely, inhale to fill my lungs, take the right end of the tube in my mouth, and blow into it as hard as I can. My cheeks sink in, my eyes bug out, and the juice inches up in the left side of the U tube, and just barely rises by—could you guess?—50 centimeters. It takes everything I have to get it there, and I can't hold it for more than a few seconds. So, I have pushed the juice up 50 centimeters on the left side, which means that I have also pushed it down 50 centimeters on the right side—in total, I have displaced the column of juice about 100 centimeters vertically, or one full meter (39 inches). Of course we are sucking air in when we breathe through a snorkel, not blowing it out. So perhaps it's easier to suck the air in? So, I do the experiment again, but this time I suck in the juice as far up the tube as I can. The result, however, is roughly the same; it only rises about 50 centimeters on the side that I suck—thus it goes down 50 centimeters on the other side, and I am utterly exhausted.

I have just imitated snorkeling 1 meter under water, the equivalent of one-tenth of an atmosphere. My students are invariably surprised by the demonstration, and they figure they can do better than their aging professor. So I invite a big strong guy to come up and give it a try, and after his best effort, his face is bright red, and he's shocked. He's only been able to do a little bit better—a couple of centimeters more—than I could.

This, it turns out, is just about the upper limit of how far down we can go and still breathe through a snorkel—1 lousy meter (about 3 feet). And we could really only manage this for a few seconds. That's why most snorkels are much shorter than 1 meter, usually only about a foot long. Try making yourself a longer snorkel—you can do so with any kind of tubing—and see what happens.

You may wonder just how much force is exerted on your chest when you submerge to do a little snorkeling. At 1 meter below the surface, the hydrostatic pressure amounts to about one-tenth of an atmosphere, or we could say one-tenth of a kilogram per square centimeter. Now the surface area of your chest is roughly one square foot, about 1,000 square centimeters. Thus the force on your chest is about 1,100 kilograms, and the force on the inner wall of your chest due to the air pressure in your lungs is about 1,000 kilograms. Therefore the one-tenth of pressure difference translates into a difference in force of 100 kilograms—a 200-pound weight! When you look at it from this perspective, snorkeling looks a lot harder, right? And if you went down 10 meters, the hydrostatic pressure would be 1 full atmosphere, 1 kilogram per square centimeter of surface, and the force on your poor chest would be about 1,000 kilograms (1 ton) higher than the outward force produced by the 1-atmosphere pressure in your lungs.

This is why Asian pearl divers—some of whom routinely dove down 30 meters—risked their lives at such depths. Because they could not snorkel, they had to hold their breath, which they could do only for a few minutes, so they had to be quick about their work.

Only now can you really appreciate the engineering achievement represented by a submarine. Let's think about a submarine at 10 meters down and assume that the air pressure inside is 1 atmosphere. The hydrostatic pressure (which is the pressure *difference* between outside and inside the sub) is about 10,000 kilograms per square meter, about 10 tons per square meter, so you can see that even a very small submarine has to be very strong to dive only 10 meters.

This is what makes the accomplishment of the fellow who invented the submarine in the early seventeenth century—Cornelis van Drebbel, who was Dutch, I'm happy to say—so astonishing. He could only operate it about 5 meters below the surface of the water, but even so, he had to deal with a hydrostatic pressure of half an atmosphere, and he built it of leather and wood! Accounts from the time say that he successfully maneuvered one of his crafts at this depth in trials on the Thames River, in England. This model was said to be powered by six oarsmen, could carry sixteen passengers, and could stay submerged for several hours. Floats held the "snorkels" just above the surface of the water. The inventor was hoping to impress King James I, trying to entice him to order a number of the crafts for his navy, but alas, the king and his admirals were not sufficiently impressed, and the sub was never used in combat. As a secret weapon, perhaps, van Drebbel's sub was underwhelming, but as a feat of engineering it was absolutely remarkable. You can find out more about Van Drebbel and early submarines at this website: www.dutch submarines.com/specials/special_drebbel.htm.

Just how far down modern navy submarines can dive is a military secret, but the prevailing wisdom is that they can go about 1,000 meters (3300 feet) deep, where the hydrostatic pressure is around 100 atmospheres, a million kilos (1,000 tons) per square meter. Not surprisingly, U.S. subs are made of very high grade steel. Russian submarines are said to be able to go even deeper, because they're made of stronger titanium.

It's easy to demonstrate what would happen to a submarine if its walls weren't strong enough, or if it dove too deep. To do this I hook up a vacuum pump to a gallon-size paint can and slowly pump the air out of the can. The pressure difference between the air outside and inside can only be as high as 1 atmosphere (compare that with the submarine!). We know that paint cans are fairly strong, but right before our eyes, because of the pressure difference, this one crumples like a flimsy aluminum soda can. It appears as though an invisible giant has taken hold of it and squeezed it in its fist. We've probably all done essentially the same thing at some point with a plastic water bottle, sucking a good bit of the air out of it and making it crumple. Intuitively, you may think the bottle scrunches up because of the power with which you've sucked on the bottle. But the real reason is that when I empty the air from the paint can, or you suck some of the air out of the water bottle, the outside air pressure no longer has enough competing pressure to push back against it. That's what the pressure of our own atmosphere is ready to do at any moment. Absolutely any moment.

A metal paint can, a plastic water bottle—these are totally mundane things, right? But if we look at them the way a physicist does, we see something entirely different: a balance of fantastically powerful forces. Our lives would not be possible without these balances of largely invisible forces, forces due to atmospheric and hydrostatic pressure, and the inexorable force of gravity. These forces are so powerful that if—or when—they get even a little bit out of equilibrium, they can cause catastrophe. Suppose a leak develops in the seam of an airplane fuselage at 35,000 feet (where the atmospheric pressure is only about 0.25 atmospheres) while the plane is traveling at 550 miles per hour? Or a hairline crack opens up in the roof of the Baltimore Harbor Tunnel, 50 feet to 100 feet below the surface of the Patapsco River?

The next time you walk down a city street, try thinking like a physicist. What are you really seeing? For one, you are seeing the result of a furious battle raging inside every single building, and I don't mean office politics. On one side of the battlefield, the force of Earth's gravitational attraction is striving to pull all of it down—not only the walls and floors and ceilings, but the desks, air-conditioning ducts, mail chutes, elevators, secretaries and CEOs alike, even the morning coffee and croissants. On the other side, the combined force of the steel and brick and concrete and ultimately the ground itself are pushing the building up into the sky.

One way to think of architecture and construction engineering, then, is that they are the arts of battling the downward force to a standstill. We may think of certain feathery skyscrapers as having escaped gravity. They've done no such thing—they've taken the battle literally to new heights. If you think about it for a little while, you'll see that the stalemate is only temporary. Building materials corrode, weaken, and decay, while the forces of our natural world are relentless. It's only a matter of time.

These balancing acts may be most threatening in big cities. Consider a horrible accident that happened in New York City in 2007, when an eighty-three-year-old 2-foot-wide pipe beneath the street suddenly could no longer contain the highly pressurized steam it carried. The resulting geyser blew a 20-foot hole in Lexington Avenue, engulfing a tow truck, and shot up higher than the nearby seventy-seven-story Chrysler Building. If such potentially destructive forces were not held in exquisite balance nearly all of the time, no one would walk any city streets.

These stalemates between immensely powerful forces are not all the product of human handiwork. Consider trees. Calm, silent, immobile, slow, uncomplaining—they employ dozens of biological strategies to combat the force of gravity as well as hydrostatic pressure. What an achievement to sprout new branches every year, to continue putting new rings on its trunk, making the tree stronger even as the gravitational attraction between the tree and the earth grows more powerful. And still a tree pushes sap up into its very highest branches. Isn't it amazing that trees can be taller than about 10 meters? After all, water can only rise

10 meters in my straw, never higher; why (and how) would water be able to rise much higher in trees? The tallest redwoods are more than 300 feet tall, and somehow they pull water all the way up to their topmost leaves.

This is why I feel such sympathy for a great tree broken after a storm. Fierce winds, or ice and heavy snow accumulating on its branches, have managed to upset the delicate balance of forces the tree had been orchestrating. Thinking about this unending battle, I find myself appreciating all the more that ancient day when our ancestors stood on two legs rather than four and began to rise to the occasion.

Bernoulli and Beyond

There may be no more awe-inspiring human achievement in defying the incessant pull of gravity and mastering the shifting winds of air pressure than flight. How does it work? You may have heard that it has to do with Bernoulli's principle and air flowing under and over the wings. This principle is named for the mathematician Daniel Bernoulli who published what we now call Bernoulli's equation in his book *Hydrodynamica* in 1738. Simply put, the principle says that for liquid and gas flows, when the speed of a flow increases, the pressure in the flow decreases. That is hard to wrap your mind around, but you can see this in action.

Hold a single sheet of paper, say an 8.5 × 11–inch standard sheet, up to your mouth (not in your mouth) with the short edge at your mouth. The paper will curl down because of gravity. Now blow hard straight out across the top of the paper, and watch what happens. You'll see the paper rise. And depending on how hard you blow, you can make the paper really jump up. You've just demonstrated Bernoulli's principle, and this simple phenomenon also helps explain how airplanes fly. Though many of us may have become used to the sight, watching a 747 take off, or being strapped in a seat when the thing lifts off, is a truly strange experience. Just watch the delight with which little children see their first plane take off. A Boeing 747-8 has a maximum takeoff weight of nearly a million pounds. How on earth does it stay aloft?

An airplane wing is designed so that the air that passes above it speeds up relative to the air that passes underneath it. Because of Bernoulli, the faster airflow on top of the wing lowers the air pressure above the wing, and the resulting difference between that low pressure and the higher pressure under the wing provides upward lift. Let's call this Bernoulli lift. Many physics books tell you that Bernoulli lift is entirely responsible for the upward lift of airplanes—in fact, this idea is all over the place. And yet, if you think about it for a minute or two, you can figure out that it cannot be true. Because if it were true, how would planes ever fly upside down?

So it's obvious that Bernoulli's principle alone cannot be the sole reason for the upward lift. In addition to the Bernoulli lift there is a so-called reaction lift. B. C. Johnson describes this in detail in his delightful article "Aerodynamic Lift, Bernoulli Effect, Reaction Lift" (http://mb-soft .com/public2/lift.html). Reaction lift (named for Newton's third law: for every action there is an equal and opposite reaction) comes about when air passes underneath an airplane wing angled upward. That air, moving from the front of the wing to the back, is pushed downward by the wing. That's the "action." That action must be met by an equal reaction of air pushing upward, so there is upward lift on the wing. In the case of a Boeing 747 (cruising at 550 miles per hour at an altitude of about 30,000 feet) more than 80 percent of the lift comes from reaction lift, and less than 20 percent from Bernoulli lift.

You can demonstrate reaction lift pretty easily yourself the next time you travel in a car. In fact, you may even have done this when you were little. When the car is moving, roll down the window, stick your arm outside, hold your hand in the direction that the car is moving, and tilt the angle of your hand such that your fingers are pointing upward. You will feel your hand pushed upward. Voila! Reaction lift.

You may think now that you understand why some planes can fly upside down. However, do you realize that if a plane rolls over 180 degrees that both the Bernoulli force and the reaction force will now be pointing downward? Remember, in normal flight the reaction force is upward because the wings are angled upward, but after a 180-degree rollover, they will be angled downward.

Do the experiment again to feel the reaction lift on your hand. As long as you tilt your fingers upward you will feel an upward force. Now change the angle such that your fingers are tilted downward; you will now feel a force in the downward direction.

Why then is it possible to fly upside down? The required lift must somehow come from an upward reaction force, since that's the only game in town. This becomes possible if the pilot (flying upside down) raises the front end of the plane enough so that the wings become angled upward again. This is a tricky business and only very experienced pilots can do it. It's also rather dangerous to rely solely on reaction lift, since by nature reaction lift is not very stable. You can sense this instability doing the experiment with your hand outside the car window. Your hand jiggles around quite a bit. In fact, it's this difficulty in controlling reaction lift that accounts for why most airplane crashes occur close to takeoff and landing. The fraction of lift accounted for by reaction lift is higher at takeoff and landing than during flight at normal altitude. This is why when a big airliner lands, you can sometimes feel the plane wobble.

The Drink Thief

The mysteries of pressure are in truth almost endlessly perplexing. Come back, for example, to the physics of drinking with a straw. Here is one last puzzle to consider, a wonderful brainteaser.

At home one weekend I said to myself, "I wonder what would be the longest straw that I could drink a glass of juice from." We've all seen super-long straws, often with turns and twists in them, which children love.

We saw earlier that we can only suck hard enough to displace juice about a maximum of 1 meter—and that only for a few seconds—meaning that I would not be able to suck up juice with a straw any higher than 1 meter (about 3 feet). So I decided to cut myself a piece of thin plastic tube 1 meter long and see if that would work. No problem; I could suck the juice up just fine. So I decided to cut a piece 3 meters long—that's almost 10 feet—and I got up on a chair in my kitchen and put a bucket of water on the floor, and sure enough, I could suck it up that far too. Amazing. Then I thought to myself, if I were up on the second story of my house and I looked down at someone below, say out on a deck having a great big tumbler of juice, wine, or whatever—let's say a *very* large cranberry and vodka—could I steal that drink by sucking it up if I had a really long straw? I decided to find out, and this led to one of the demonstrations I love to do in class. It never ceases to amaze the students.

I pull out a long length of coiled-up clear plastic tubing and I ask for a front-row volunteer. I place a large glass beaker of cranberry juice—no vodka—on the floor in the classroom for all students to see. Holding the tubing, I begin to climb a tall ladder; it reaches about 16 feet off the floor—almost 5 meters!

"Okay, here's my straw," I say, dropping one end of the tubing to the student. She holds the end in the beaker, and I can feel the students' anticipation. The class can't quite believe I'm up there. Remember, they were witnesses to the fact that I could only displace the cranberry juice about 1 meter, or about 3 feet. Now I'm about 16 feet off the ground. How could I possibly do it?

I begin sucking, grunting a bit as the juice rises slowly inside the tube: first 1 meter, then 2, and then 3. Then the level dips a little, but soon the juice resumes climbing very slowly again until it reaches my mouth. I say a loud "Mmmmm" and the class erupts in applause. What has been going on here? Why could I suck the juice up so high?

Frankly, I cheat. Not that it matters, since there are no rules in the game. Every time after sucking, when I can't take any more air in, I put my tongue over the end of the tube. In other words I close the tube off, and as we saw earlier, this will keep the juice up in the tube. I then exhale and I start sucking again, and repeat that scenario many times. My mouth becomes a kind of suction pump and my tongue a kind of stop valve.

To make the juice rise those 16 feet, I have to lower the pressure of the air in the tube to about half an atmosphere. And yes, if you're wondering, I could have used the same trick with the manometer, and I would have been able to suck up a much longer column of cranberry juice. Does that mean that I could also snorkel much farther down beneath the surface of a lake or the sea?

What do you think? If you know the answer, drop me a note!

CHAPTER 5

Over and Under—Outside and Inside— the Rainbow

S o many of the little wonders of the everyday world—which can be truly spectacular—go unobserved most of the time because we haven't been trained how to see them. I remember one morning, four or five years ago, when I was drinking my morning espresso in my favorite red and blue Rietveld chair, and suddenly I noticed this beautiful pattern of round spots of light on the wall, amidst the flickering of shadows thrown by the leaves of a tree outside the window. I was so delighted to have spotted them that my eyes lit up. Not sure what had happened, but with her usual astuteness, my wife, Susan, wondered if something was the matter.

"Do you know what that is?" I responded, pointing to the light circles. "Do you understand why that's happening?" Then I explained. You might expect the light to make lots of little shimmerings on the wall rather than circles, right? But each of the many small openings between the leaves was acting like a camera obscura, a pinhole camera, and such a camera reproduces the image of the light source—in this case the Sun. No matter what the shapes of the openings through which the light is streaming, as long as the opening is small, it's the shape of the light source itself that's re-created on the wall.

So during a partial solar eclipse, sunlight pouring through my window wouldn't make circles on my wall anymore—all the circles would have a bite taken out of them, because that would be the shape of the Sun. Aristotle knew this more than two thousand years ago! It was fantastic to see those light spots, right there on my bedroom wall, demonstrating the remarkable properties of light.

Secrets of the Rainbow

In truth, the marvelous effects of the physics of light are everywhere we look, sometimes in the most ordinary sights, and at other times in some of nature's most beautiful creations. Take rainbows, for example: fantastic, wonderful phenomena. And they're everywhere. Great scientists—Ibn al-Haytham, the eleventh-century Muslim scientist and mathematician known as the father of optics, the French philosopher, mathematician, and physicist René Descartes; and Sir Isaac Newton himself—found them captivating and tried to explain them. Yet most physics teachers ignore rainbows in their classes. I can't believe this; in fact, I think it's *criminal*.

Not that the physics of rainbows is simple. But so what? How can we refuse to tackle something that pulls so powerfully on our imaginations? How can we not want to understand the mystery behind the intrinsic beauty of these glorious creations? I have always loved lecturing about them, and I tell my students, "At the end of this lecture, your life will never be the same again, never." The same goes for you.

Former students and others who've watched my lectures on the web have been mailing and emailing me wonderful images of rainbows and other atmospheric phenomena for decades. I feel as though I have a network of rainbow scouts spread across the world. Some of these shots are extraordinary—especially those from Niagara Falls, which has so much spray that the bows are spectacular. Maybe you will want to send me pictures too. Feel free!

I'm sure you've seen at least dozens, if not hundreds, of rainbows in your life. If you've spent time in Florida or Hawaii, or other tropical locations where there are frequent rain showers while the Sun shines, you've seen even more. If you've watered your garden with a hose or sprinkler when the Sun is shining, you've probably created rainbows.

Most of us have *looked at* many rainbows, yet very few of us have ever *seen* rainbows. Ancient mythologies have called them gods' bows, bridges or paths between the homes of mortals and the gods. Most famously in the West, the rainbow represented God's promise in the Hebrew Bible never again to bring an all-destroying flood to the earth: "I do set my bow in the clouds."

Part of the charm of rainbows is that they are so expansive, spreading majestically, and so ephemerally, across the entire sky. But, as is so often true in physics, their origins lie in extraordinarily large numbers of something exceptionally minute: tiny spherical balls of water, sometimes less than 1 millimeter (½₅ of an inch) across, floating in the sky.

While scientists have been trying to explain the origins of rainbows for at least a millennium, it was Isaac Newton who offered the first truly convincing explanation in his 1704 work *Opticks*. Newton understood several things at once, all of which are essential for producing rainbows. First, he demonstrated that normal white light was composed of all the colors (I was going to say of "all the colors of the rainbow," but that would be getting ahead of ourselves). By refracting (bending) light through a glass prism, he separated it into its component colors. Then, by sending the refracted light back through another prism, he combined the colored light back into white light, proving that the prism itself wasn't creating the colors in some way. He also figured out that many different materials could refract light, including water. And this is how he came to understand that raindrops refracting and reflecting light were the key to producing a rainbow.

A rainbow in the sky, Newton concluded correctly, is a successful collaboration between the Sun, zillions of raindrops, and your eyes, which must be observing those raindrops at just the right angles. In order to understand just how a rainbow is produced, we need to zero in on what happens when light enters a raindrop. But remember, everything I'm going to say about this single raindrop in reality applies to the countless drops that make up the rainbow.

For you to see a rainbow, three conditions need to be met. First, the Sun needs to be behind you. Second, there must be raindrops in the sky in front of you—this could be miles or just a few hundred yards away. Third, the sunlight must be able to reach the raindrops without any obstruction, such as clouds.

When a ray of light enters a raindrop and refracts, it separates into all of its component colors. Red light refracts, or bends, the least, while violet light refracts the most. All of these different-colored rays continue traveling toward the back of the raindrop. Some of the light keeps going and exits the raindrop, but some of it bounces back, or reflects, at an angle, toward the front of the raindrop. In fact, some of the light reflects more than once, but that only becomes important later. For the time being, we are only interested in the light that reflects just once. When the light exits the front of the drop, some of the light again refracts, separating the different colored rays still further.

After these rays of sunlight refract, reflect, and refract again on their way out of the raindrop, they have pretty much reversed direction. Key to why we see rainbows is that red light exits the raindrop at angles that are *always smaller* than about 42 degrees from the original direction of the sunlight entering the drop. And this is the same for all raindrops, because the Sun for all practical purposes is infinitely far away. The angle at which the red light exits can be anything between 0 degrees and 42 degrees but never more than 42 degrees, and this maximum angle is different for each of the different colors. For violet light, the maximum angle is about 40 degrees. These different maximum angles for each color account for the stripes of colors in the rainbow.

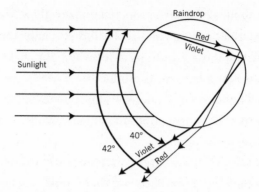

There is an easy way to spot a rainbow when conditions are right. As seen in the following figure, if I trace a line from the Sun through my head to the far end of my shadow on the ground, that line is precisely parallel to the direction from the Sun to the raindrops. The higher the Sun in the sky, the steeper this line will be, and the shorter my shadow. The converse is also the case. This line, from the Sun, through my head, to the shadow of my head on the ground, we will call the imaginary line. This line is very important as it will tell you where in the sky you should look to see the rainbow.

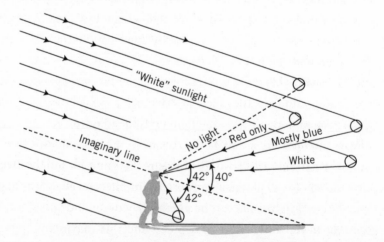

All raindrops at 42 degrees from the "imaginary line" will be red. Those at 40 degrees will be blue. Raindrops at angles smaller than 40 degrees will be white (like the sunlight). We will see no light from drops at angles larger than 42 degrees (see text).

If you look about 42 degrees away from that imaginary line—it doesn't matter whether it's straight up, to the right, or to the left—that's where you will see the red band of the rainbow. At about 40 degrees away from it—up, right, or left—you will see the violet band of the rainbow. But the truth is that violet is hard to see in a rainbow, so you'll see the blue much more easily. Therefore we'll just say blue from now on. Aren't these the same angles I mentioned earlier, talking about the maximum angles of the light leaving the raindrop? Yes, and it's no accident. Look again at the figure.

What about the blue band in the rainbow? Well, remember its magic number is just about 40 degrees, 2 degrees less than the red band. So blue light can be found refracting, reflecting, and refracting out of *different raindrops* at a *maximum* angle of 40 degrees. Thus we see blue light 40 degrees away from the imaginary line. Since the 40-degree band is closer to the imaginary line than the 42-degree band, the blue band will always be on the *inside* of the red band of the rainbow. The other colors making up the bow—orange, yellow, green—are found between the red and blue bands. For more about this you can take a look at my lecture on rainbows online, at http://ocw.mit.edu/courses/physics/8-03-physics-iii -vibrations-and-waves-fall-2004/video-lectures/lecture-22/.

Now you might wonder, at the maximum angle for blue light, are we seeing only blue light? After all, red light can also emerge at 40 degrees, as it is smaller than 42 degrees. If you've asked this question, more power to you—it's a very astute one. The answer is that at the maximum angle for any given color, that color dominates all other colors. With red, though, because its angle is the highest, it *is* the only color.

Why is the rainbow a bow and not a straight line? Go back to that imaginary line from your eyes to the shadow of your head, and the magic number 42 degrees. When you measure 42 degrees—in all directions— away from the imaginary line, you are tracing an arc of color. But you know that not all rainbows are full arcs—some are just little pieces in the sky. That happens when there aren't enough raindrops in all directions in the sky or when certain parts of the rainbow are in the shadow of obstructing clouds.

There's another important aspect to this collaboration between the Sun, the raindrops, and your eyes, and once you see it, you'll understand lots more about why rainbows—natural as well as artificial—are the way they are. For example, why are some rainbows enormous, while others just hug the horizon? And what accounts for the rainbows you sometimes see in pounding surf, or in fountains, waterfalls, or the spray of your garden hose?

Let's go back to the imaginary line that runs from your eyes to the shadow of your head. This line starts at the Sun, behind you, and extends to the ground. However, in your mind, you can extend this line as far as you want, even much farther than the shadow of your head. This imaginary line is very useful, as you can imagine it going through the center (called the antisolar point) of a circle, on the circumference of which is the rainbow. This circle represents where the rainbow would form if the surface of Earth didn't get in its way. Depending upon how high in the sky the Sun is, a rainbow will also be lower or higher above the horizon. When the Sun is very high, the rainbow may only just peek above the horizon, whereas late in the afternoon just before sunset, or early in the morning just around sunrise, when the Sun is low in the sky and when your shadow is long, then a rainbow may be enormous, reaching halfway up into the sky. Why halfway? Because the maximum angle it can be over the horizon is 42 degrees, or close to 45 degrees, which is half of the 90 degrees that would be right overhead.

So how can you go rainbow hunting? First of all, trust your instincts about when a rainbow might form. Most of us tend to have a good intuitive sense of that: those times when the Sun is shining just before a rainstorm, or when it comes out right after one. Or when there's a light shower and the sunlight can still reach the raindrops.

When you feel one coming on, here's what you do. First, turn your back to the Sun. Then locate the shadow of your head, and look about 42 degrees in any direction away from the imaginary line. If there's enough sunlight, and if there are enough raindrops, the collaboration will work and you will see a colorful bow.

Suppose you cannot see the Sun at all—it's somehow hidden by clouds or buildings, but it's clearly shining. As long as there are no clouds between the Sun and the raindrops, you still ought to be able to see a rainbow. I can see rainbows in the late afternoon from my living room facing east when I cannot see the Sun that is in the west. Indeed, most of the time you don't need the imaginary line and the 42-degree trick to spot a rainbow, but there is one situation where paying attention to both can make a big difference. I love to walk on the beaches of Plum Island off the Massachusetts coast. Late in the afternoon the sun is in the west and the ocean is to the east. If the waves are high enough and if they make lots of small water drops, these droplets act like raindrops and you can see two small pieces of the rainbow: one piece at about 42 degrees to the left of the imaginary line and a second piece about 42 degrees to the right. These rainbows only last for a split second, so it's a huge help in spotting them if you know where to look in advance. Since there are always more waves coming, you will always succeed if you can be patient enough. More about this later in this chapter.

Here is another thing you can try to look for, the next time you spot a rainbow. Remember our discussion of the maximum angle at which certain light can refract out of the raindrop? Well, even though you will see blue, or red, or green from certain raindrops, raindrops themselves cannot be so choosy: they refract, reflect, and refract lots of light at *less* than a 40-degree angle too. This light is a mixture of all the different colors at roughly equal intensities, which we see as white light. That's why, inside the blue band of a rainbow, the sky is very bright and white. At the same time, *none* of the light that refracts, reflects, and refracts again can exit raindrops beyond the 42-degree angle, so the sky just outside the bow is darker than inside the bow. This effect is most visible if you compare the brightness of the sky on either side of the rainbow. If you're not specifically looking for it, you probably won't even notice it. There are excellent images of rainbows in which you can see this effect on the Atmospheric Optics website, at www.atoptics.co.uk.

Once I began explaining rainbows to my students, I realized just how rich a subject they are—and how much more I had to learn. Take double rainbows, which you've probably seen from time to time. In fact, there are almost always two rainbows in the sky: the so-called primary bow, the one I've been discussing, and what we call the secondary bow.

If you've seen a double rainbow, you've probably noticed that the secondary bow is much fainter than the primary bow. You probably *haven't* noticed, though, that the order of colors in the secondary bow is blue on the outside and red on the inside, the reverse of that in the primary. There is an excellent photograph of a double rainbow in this book's photo insert.

In order to understand the origin of the secondary bow, we have to go back to our ideal raindrop—remember, of course, that it actually takes zillions of drops to make up a secondary rainbow as well. Some of the light rays entering the drops reflect just once; others reflect twice before exiting. While light rays entering any given raindrop can reflect many times inside it, the primary bow is only created by those that reflect *once*. The secondary bow, on the other hand, is created only by those that reflect *twice* inside before refracting on the way out. This extra bounce inside the raindrop is the reason the colors are reversed in the secondary bow.

The reason the secondary bow is in a different position from the primary bow—always outside it—is that twice-reflected red rays exit the drop at angles always *larger* (yes, larger) than about 50 degrees, and the twice-reflected blue rays come out at angles always larger than about 53 degrees. You therefore need to look for the secondary bow about 10 degrees *outside* the primary bow. The reason that the secondary bow is much fainter is that so much less light reflects inside the raindrops twice than reflects once, so there's less light to make the bow. This is, of course, why it can be hard to see the secondary bow, but now that you know they often accompany primary rainbows, and where to look for them, I'm confident you'll see lots more. I also suggest that you spend a few minutes on the Atmospheric Optics website.

Now that you know what makes rainbows, you can perform a little optical magic in your own backyard or on your driveway or even on the sidewalk, with just a garden hose. But because you can manipulate the raindrops, and they are physically close to you, there are a couple of big differences. For one thing, you can make a rainbow even when the Sun is high in the sky. Why? Because you can make raindrops between you and your shadow on the ground, something that rarely happens naturally. As long as there are raindrops that the sunlight can reach, there can be rainbows. You may have done this already, but perhaps not as purposefully.

If you have a nozzle on the end of the hose, adjust it to a fine spray, so the droplets are quite small, and when the Sun is high in the sky, point the nozzle toward the ground and start spraying. You cannot see the entire circle all at once, but you will see pieces of the rainbow. As you continue moving the nozzle in a circle, piece by piece you will see the entire circle of the rainbow. Why do you have to do it this way? Because you don't have eyes in the back of your head!

You will see red at about 42 degrees from the imaginary line, the inside edge of the circular bow will be blue, and inside the bow you will see white light. I love performing this little act of creation while watering my garden, and it's especially satisfying to be able to turn all the way around and make a complete 360-degree rainbow. (The Sun, of course, will then not always be behind you.)

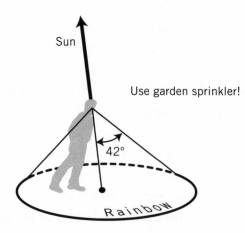

One cold winter day in 1972 I was so determined to get some good photos of these homemade rainbows for my class that I made my poor daughter Emma, who was just seven, hold the hose in my yard, squirting the water high in the air, while I snapped away with the camera. But I guess when you're the daughter of a scientist you have to suffer a little bit for the sake of science. And I did get some great pictures; I even managed to photograph the secondary bow, using my contrasting blacktop driveway as the background. You can see the picture of Emma in the insert.

I hope you'll try this experiment—but do it in the summer. And don't be too disappointed if you don't see the secondary bow—it may be too faint to show up if your driveway isn't dark enough.

From now on, with this understanding of how to spot rainbows, you'll find yourself compelled to look for them more and more. I often can't help myself. The other day as Susan and I were driving home, it started to rain, but we were driving directly west, into the Sun. So I pulled over, even though there was a good deal of traffic; I got out of the car and turned around, and there it was, a real beauty!

I confess that whenever I walk by a fountain when the sun is shining, I position myself so I can search for the rainbow I know will be there. If you're passing by a fountain on a sunny day, give it a try. Stand between the Sun and the fountain with your back to the Sun, and remember that the spray of a fountain works just like raindrops suspended in the sky. Find the shadow of your head—that establishes the imaginary line. Then look 42 degrees away from that line. If there are enough raindrops in that direction, you'll spy the red band of the rainbow and then the rest of the bow will come immediately into view. It's rare that you see a full semicircular arc in a fountain—the only way you can see one is to be very close to the fountain—but the sight is so beautiful, it's always worth trying.

Once you've found it, I warn you that you may just feel the urge to let your fellow pedestrians know it's there. I often point these fountain rainbows out to passersby, and I'm sure some of them think I'm weird. But as far as I'm concerned, why should I be the only one to enjoy such hidden wonders? Of *course* I show them to people. If you know a rainbow could be right in front of you, why not look for it, and why not make sure others see it too? They are just so beautiful.

Students often ask me whether there is also a tertiary bow. The answer is yes and no. The tertiary bow results, as you might have guessed, from three reflections inside the raindrop. This bow is centered on the Sun and, like the primary bow, which is centered on the antisolar point, it also has a radius of about 42 degrees and red is on the outside. Thus you have to look toward the Sun to see it and it has to rain between you and the Sun. But when that is the case, you will almost never see the Sun. There are additional problems: a lot of sunlight will go through the raindrops without reflecting at all and that produces a very bright and very large glow around the Sun which makes it effectively impossible to see the tertiary bow. The tertiary bow is even fainter than the secondary. It is also much broader than the primary and the secondary bow; thus the already faint light of the bow is spread out even more over the sky and that makes it even more difficult to see it. As far as I know, no pictures of tertiary bows exist, and I do not know of anyone who has ever seen a tertiary bow. Yet there are some reports of sightings.

Inevitably, people want to know if rainbows are real. Maybe they're mirages, they wonder, receding endlessly as we approach them. After all, why can't we see the end of the rainbow? If this thought has been at the back of your mind, breathe easy. Rainbows are real, the result of real sunlight interacting with real raindrops and your real eyes. But since they result from a precise collaboration between your eyes, the Sun, and the raindrops, you will see a different rainbow from the person across the street. Equally real, but different.

The reasons we usually cannot see the end of the rainbow touching the Earth are not because it doesn't exist, but because it's too far away, or hidden by buildings or trees or mountains, or because there are fewer raindrops in the air there and the bow is too faint. But if you can get close enough to a rainbow, you can even touch it, which you should be able to do with the rainbow you make with your garden hose.

I have even taken to holding rainbows in my hand while I shower. I accidentally discovered this one day. When I faced the shower spray, I suddenly saw two (yes two!) bright primary rainbows inside my shower, each one about a foot long and an inch wide. This was so exciting, so beautiful; it was like a dream. I reached out and held them in my hands. Such a feeling! I'd been lecturing on rainbows for forty years, and never before had I seen two primary rainbows within arm's reach.

Here's what had happened. A sliver of sunlight had shone into my shower through the bathroom window. In a way, it was as though I was standing not in front of a fountain, but inside the fountain. Since the water was so close to me and since my eyes are about three inches apart, each eye had its own, distinct imaginary line. The angles were just right, the amount of water was just right, and each of my eyes saw its own primary rainbow. When I closed one eye, one of the rainbows would disappear; when I closed the other eye, the other bow vanished. I would have loved to photograph this astonishing sight, but I couldn't because my camera has only one "eye."

Being so close to those rainbows that day made me appreciate in a new way just how real they were. When I moved my head, they too moved, but as long as my head stayed where it was, so did they.

Occasionally I time my morning showers whenever possible to catch these rainbows. The Sun has to be at the right location in the sky to peek through my bathroom window at the right angle and this only happens between mid-May and mid-July. You probably know that the Sun rises earlier and goes higher in the sky in certain months, and that in the Northern Hemisphere it rises more to the south (of east) than in the winter months, and more to the north (of east) in summer.

My bathroom window faces south, and there's a building on the south side, making sure that light can never enter from due south. So sunlight only comes in roughly from the southeast. The time I first saw the shower bows was while I was taking a very late shower, around ten o'clock. In order to see rainbows in your own shower you will need a bathroom window through which sunlight can reach the spray. In fact, if you can never see the Sun by looking out your bathroom window, there's no point in looking for shower bows—there just won't be any. The sunlight must be able to actually reach your shower. And even if it does come directly in, that's no guarantee, because many water drops have to be present at 42 degrees from your imaginary line, and that may not be the case.

These are probably difficult conditions to meet, but why not try? And if you discover that the Sun enters your shower just right late in the afternoon, well, then, you could always think about changing your shower schedule.

Why Sailors Wear Sunglasses

Whenever you do decide to go rainbow hunting, be sure to take off your sunglasses if they are the kind we call polarized, or you might miss out on the show. I had a funny experience with this one day. As I said, I love to take walks along the beaches of Plum Island. And I've explained how you can see little bows in the spray of the waves. Years ago I was walking along the beach. The sun was bright and the wind was blowing, and when the waves rolled over as they got close to the beach, there was lots of spray—so I was frequently seeing small pieces of bows as I mentioned earlier in this chapter. I started pointing them out to my friend, who said he couldn't see what I was talking about. We must have gone back and forth half a dozen times like this. "There's one," I would shout, somewhat annoyed. "I don't see anything!" he would shout back. But then I had a bright moment and I asked him to take off his sunglasses, which I looked at—sure enough, they were polarized sunglasses. Without his sunglasses he did see the bows, and he even started to point them out to me! What was going on?

Rainbows are something of an oddity in nature because almost all of their light is polarized. Now you probably know the term "polarized" as a description of sunglasses. The term is not quite technically correct, but let me explain about polarized light—then we'll get to the sunglasses and rainbows.

Waves are produced by vibrations of "something." A vibrating tuning fork or violin string produces sound waves, which I talk about in the next chapter. Light waves are produced by vibrating electrons. Now, when the vibrations are all in one direction and are perpendicular to the direction of the wave's propagation, we call the waves linearly polarized. For simplicity I will drop the term "linearly" in what follows as I am only talking in this chapter about this kind of polarized light.

Sound waves can never be polarized, because they always propagate in the same direction as the oscillating air molecules in the pressure waves; like the waves you can generate in a Slinky. Light, however, can be polarized. Sunlight or light from lightbulbs in your home is not polarized. However, we can easily convert nonpolarized light into polarized light. One way is to buy what are known as polarized sunglasses. You now know why their name isn't quite right. They are really polarizing sunglasses. Another is to buy a linear polarizer (invented by Edwin Land, founder of the Polaroid Corporation) and look at the world through it. Land's polarizers are typically 1 millimeter thick and they come in all sizes. Almost all the light that passes through such a polarizer (including polarizing sunglasses) has become polarized.

If you put two rectangular polarizers on top of each other (I hand out two of them to each of my students, so they can experiment with them at home) and you turn them 90 degrees to each other, no light will pass through.

Nature produces lots of polarized light without the help of one of Land's polarizers. Light from the blue sky 90 degrees away from the direction of the Sun is nearly completely polarized. How can we tell? You look at the blue sky (anywhere at 90 degrees away from the Sun) through one linear polarizer and rotate it slowly while looking through it. You will notice that the brightness of the sky will change. When the sky becomes almost completely dark, the light from that part of the sky is nearly completely polarized. Thus, to recognize polarized light, all you need is one polarizer (but it's much more fun to have two).

In the first chapter I described how in class I "create" blue light by scattering white light off cigarette smoke. I arrange this in such a way that the blue light that scatters into the lecture hall has scattered over an angle of about 90 degrees; it too is nearly completely polarized. The students can see this with their own polarizers, which they always bring with them to lectures.

Sunlight that has been reflected off water or glass can also become nearly completely polarized if the sunlight (or light from a lightbulb) strikes the water or glass surface at just the right angle, which we call the Brewster angle. That's why boaters and sailors wear polarizing sunglasses—they block much of the light reflecting off the water's surface. (David Brewster was a nineteenth-century Scottish physicist who did a lot of research in optics.)

I always carry at least one polarizer with me in my wallet—yes, *always*—and I urge my students to do the same.

Why am I telling you all this about polarized light? Because the light from rainbows is nearly completely polarized. The polarization occurs as the sunlight inside the water drops reflects, which, as you now know, is a necessary condition for rainbows to be formed.

I make a special kind of rainbow in my classes (using a single, though very large, water drop) and I am able to demonstrate that (1) red is on the outside of the bow, (2) blue is on the inside, (3) inside the bow the light is bright and white, which is not the case outside the bow, and (4) the light from the bow is polarized. The polarization of the bows for me is very fascinating (one reason why I always carry polarizers on me). You can see this wonderful demonstration in my lecture at http://ocw.mit .edu/courses/physics/8-03-physics-iii-vibrations-and-waves-fall-2004/ video-lectures/lecture-22/.

Beyond the Rainbow

Rainbows are the best known and most colorful atmospheric creations, but they are far from alone. There is an entire host of atmospheric phenomena, some of them really quite strange and striking, others deeply mysterious. But let's stay with rainbows for a bit and see where they take us.

When you look carefully at a very bright rainbow, on its inner edge you may sometimes see a series of alternating bright-colored and dark bands—which are called supernumerary bows. You can see one in the insert. To explain these we must abandon Newton's explanation of light rays. He thought that light was composed of particles, so when he imagined individual rays of light entering, bouncing around in, and exiting raindrops, he assumed that these rays acted as though they were little particles. But in order to explain supernumerary bows we need to think of light as consisting of waves. And in order to make a supernumerary bow, light waves must go through raindrops that are really small, smaller than a millimeter across.

One of the most important experiments in all of physics (generally referred to as the double-slit experiment) demonstrated that light is made of waves. In this famous experiment performed around 1801–03, the English scientist Thomas Young split a narrow beam of sunlight into two beams and observed on a screen a pattern (the sum of the two beams) that could only be explained if light consists of waves. Later in time this experiment was done differently using actually two slits (or two pinholes). I will proceed here assuming that a narrow beam of light strikes two very small pinholes (close together) made in a thin piece of cardboard. The light passes through the pinholes and then strikes a screen. If light was made of particles, a given particle would either go through one pinhole or through the other (it cannot go through both) and thus you would see two bright spots on the screen. However, the pattern observed is very different. It precisely mimics what you'd expect if two waves had hit the screen—one wave coming from one pinhole and simultaneously one identical wave coming from the other. Adding two waves is subject to what we call interference. When the crests of the waves from one pinhole line up with the valleys of waves from the other, the waves cancel each other, which is called destructive interference, and the locations on the screen where that happens (and there are several) remain dark. Isn't that amazing—light plus light turns into darkness! On the other hand, at other locations on the screen where the two waves are in sync with one another, cresting and falling with one another, we have constructive interference and we end up with bright spots (and there will be several). Thus we will see a *spread out* pattern on the screen consisting of alternating dark and bright spots, and that is precisely what Young observed with his split-beam experiment.

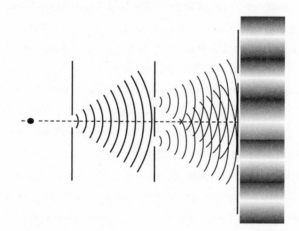

I demonstrate this in my classes using red laser light and also with green laser light. It's truly spectacular. Students notice that the pattern of the green light is very similar to that of the red light except that the separation between the dark and the bright spots is somewhat smaller for the green light. This separation depends on the color (thus wavelength) of light (more about wavelength in the next chapter).

Scientists had been battling for centuries over whether light consisted of particles or waves, and this experiment led to the stunning and indisputable conclusion that light is a wave. We now know that light can act both as a particle and as a wave, but *that* astounding conclusion had to wait another century, for the development of quantum mechanics. We don't need to go further into that at the moment.

Going back to supernumerary bows, interference of light waves is what creates the dark and bright bands. This phenomenon is very pronounced when the diameter of the drops is near 0.5 millimeters. You can see some images of supernumerary bows at www.atoptics.co.uk/rainbows/supdrsz.htm.

The effects of interference (often called diffraction) become even more dramatic when the diameters of the droplets are smaller than about 40 microns (0.04 millimeters, or 1/635 of an inch). When that happens, the colors spread out so much that the waves of different colors completely overlap; the colors mix and the rainbow becomes white. White rainbows often show one or two dark bands (supernumerary bows). They are very rare and I have never seen one. A student of mine, Carl Wales, sent me pictures in the mid-1970s of several beautiful white rainbows. He had taken the pictures in the summer at two a.m. (yes, two a.m.) from Fletcher Ice Island, which is a large drifting iceberg (about 3 × 7 miles). At the time, it was about 300 miles from the North Pole. You can see a nice picture of a white rainbow in the insert.

These white bows can also be seen in fog, which consists of exceptionally tiny water droplets. White fogbows can be hard to spot; you may have seen them many times without realizing it. They are likely to appear whenever fog is thin enough for sunlight to shine through it. When I'm on a riverbank or in a harbor in the early morning, when the Sun is low in the sky, and where fog is common, I hunt for them and I have seen many.

Sometimes you can even create a fogbow with your car headlights. If you're driving and the night fog rolls in around you, see if you can find a safe place to park. Or, if you're at home and the fog comes, face your car toward the fog and turn on your headlights. Then walk away from your car and look at the fog where your headlight beams are. If you're lucky, you might be able to see a fogbow. They make the gloom of a foggy night even spookier. You can see the results of a fellow stumbling across fogbows that he made with his car headlights at www.extremeinstability .com/08-9-9.htm. Did you notice the dark bands in the white bows?

The size of water droplets and the wave nature of light also explain another of the most beautiful phenomena that grace the skies: glories. They can best be seen when flying over clouds. Trust me, it's worth trying to find them. In order to do so, you must, of course, be in a window seat—and not over the wings, which block your view down. You want to make certain that the Sun is on the side of the plane opposite your seat, so you'll have to pay attention to the time of day and the direction of the flight. If you can see the Sun out your window, the experiment is over. (I have to ask you to trust me here; a convincing explanation requires a lot of very complicated math.) If these conditions are met, then try to figure out where the antisolar point is and look down at it. If you've hit the jackpot you may see colorful rings in the clouds and if your plane is flying not too far above the clouds, you may even see the glory circling the shadow of the plane—glories have diameters that can vary from a few degrees to about 20 degrees. The smaller the drops, the larger the glories.

I have taken many pictures of glories, including some where the shadow of my plane was clearly visible and the really fun part is that the position of my seat is at the center of the glory, which is the antisolar point. One of these pictures is in the insert.

You can find glories in all kinds of places, not just from airplanes. Hikers often see them when they have the Sun to their backs and look down into misty valleys. In these cases, a quite spooky effect happens. They see their own shadow projected onto the mist, surrounded by the glory, sometimes several colorful rings of it, and it looks positively ghostly. This phenomenon is also known as the Brocken spectre (also called Brocken bow), named for a high peak in Germany where sightings of glories are common. In fact, glories around people's shadows look so much like saintly halos, and the figures themselves look so otherworldly, that you will not be surprised to learn that *glory* is actually an old word for the circle of light around the head of various saints. In China, glories are known as Buddha's light.

I once took a marvelous photo of my own shadow surrounded by a glory that I refer to as the image of Saint Walter. A good many years ago I was invited by some of my Russian astronomer friends to their 6-meter telescope in the Caucasus Mountains. This was the world's largest telescope at the time. The weather was just awful for observing. Every day I was there, at about five thirty in the afternoon a wall of fog would come rolling up out of the valley below and completely engulf the telescope. I mean totally; we couldn't make any observations at all during my visit. A picture of the fog ascending is shown in the insert. In talking to the astronomers, I learned that the fog was very common. So I asked, "Why then was this telescope built here?" They told me that the telescope was built on that site because the wife of a Party official wanted it right there, and that was that. I almost fell off my chair.

Anyway, after a few days, I got the idea that I might be able to take a fantastic photo. The Sun was still strong in the west every day when the fog came up from the valley, which was to the east, the perfect setup for glories. So the next day I brought my camera to the observatory, and I was getting nervous that the fog might not cooperate. But sure enough, the wall of fog swelled up, and the Sun was still shining, and my back was to it. I waited and waited and then, boom, there was the glory around my shadow and I snapped. I couldn't wait to develop the film—this was in the pre-digital age—and there it was! My shadow is long and ghostly, and the shadow of my camera is at the center of the rings of a gorgeous glory. You can see the picture in the insert.

You don't need to be in such an exotic location to see a halo around your head. On a sunlit early morning if you look at your shadow on a patch of dewy grass (of course with the Sun right behind you), you can often see what in German is called *Heiligenschein*, or "holy light": a glow around the shadow of your head. (It's not multicolored; it's not a glory.) Dewdrops on the grass reflect the sunlight and create this effect. If you try this—and I hope you will—they're easier to find than glories. You will see that since it's early morning and the Sun is low, your shadow will be quite long, and you appear much like the elongated and haloed saints of medieval art.

The many different types of bows and halos can surprise you in the most unexpected places. My favorite sighting happened one sunny day in June 2004—I remember it was the summer solstice, June 21—when I was visiting the deCordova Museum in Lincoln, Massachusetts, with Susan (who was not yet my wife at the time), my son, and his girlfriend. We were walking across the grounds toward the entrance when my son called out to me. There in front of us, on the ground, was a stunning, colorful, nearly circular bow. (Because it was the solstice, the Sun was as high as it ever gets in Boston, about 70 degrees above the horizon.) It was breathtaking!

I pulled out my camera and snapped a bunch of photos as quickly as I could. How unexpected. There were no water droplets on the ground, and I quickly realized the bow could not have been made from water drops in any event because the radius of the bow was much smaller than 42 degrees. And yet it looked just like a rainbow: the red was on the outside, the blue was on the inside, and there was bright white light inside the bow. What could have caused it? I realized that it must have been made by transparent, spherical particles of something, but what could they be?

One of my photographs of the bow, which you can see in the insert, turned out so well that it became NASA's astronomical mystery picture of the day, posted on the web on September 13, 2004.* (This, by the way, is a terrific website, and you should look at it every day at http://apod .nasa.gov/apod/astropix.html.) I received about three thousand guesses as to what it was. My favorite response was a handwritten note from Benjamin Geisler, age four, who wrote, "I think your mystery photo is made by light, crayons, markers and colored pencils." It's posted on the

---

*If you want to see my photo online, click on the website's Archive and go to 2004 September 13. See text above for the general URL.

bulletin board outside my office at MIT. Of all the answers, about thirty were on the right track, but only five were dead on.

The best clue to this puzzle is that there was a good bit of construction going on at the museum when we visited. In particular, there had been a lot of sandblasting of the museum's walls. Markos Hankin, who was in charge of the physics demonstrations at MIT and with whom I have worked for many years, told me—I didn't know this at the time—that some kinds of sandblasting use glass beads. And there *were* lots of tiny glass beads lying on the ground. I had taken a few spoonfuls of the beads home. What we had seen was a glassbow, which has now become an official category of bow, a bow formed by glass beads; it has a radius of about 28 degrees, but the exact value depends on the kind of glass.

Markos and I couldn't wait to see if we could make one of our own for my lectures. We bought several pounds of glass beads, glued them on big sheets of black paper, and attached the paper to a blackboard in the lecture hall. Then, at the end of my lecture on rainbows, we aimed a spotlight on the paper from the back of the lecture hall. It worked! I invited the students to come, one by one, to the front of the class, where they stood before the blackboard and cast their shadow smack in the middle of their own private glassbow.

This was such a thrilling experience for the students that you might want to try it at home; making a glassbow is not too difficult. It does depend on what your objectives are. If you just want to see the colors of the bow, it's quite easy. If you want to see the entire bow encircling your head it's more work.

To see a small piece of the bow, all you need is a piece of black cardboard about one foot square, some clear spray adhesive (we used 3M's Spray Mount Artist's Adhesive, but any clear spray glue will do), and transparent spherical glass beads. They must be transparent and spherical. We used "coarse glass bead blast media," with diameters ranging from 150 to 250 microns, which you can find here: http://tinyurl.com/glassbeads4rainbow.

Spray glue on your cardboard, and then sprinkle the beads on it. The average distance between the beads isn't critical, but the closer the beads are, the better. Be careful with these beads—you probably want to do this outside so you don't spill beads all over your floor. Let the glue dry, and if you have a sunny day, go outside.

Establish the imaginary line (from your head to the shadow of your head). Place the cardboard somewhere on that line; thus you will see the shadow of your head on the cardboard (if the Sun is low in the sky, you could put the cardboard on a chair; if the Sun is high in the sky you could put it on the ground—remember the glass beads at the deCordova museum were also on the ground. You may select how far away from your head you place the cardboard. Let's assume that you place it 1.2 meters (about 4 feet) away. Then move the cardboard about 0.6 meters (2 feet) away from the imaginary line in a direction perpendicular to the line. You may do that in any direction (left, right, up, down)! You will then see the colors of the glassbow. If you prefer to place the cardboard farther away, say 1.5 meters (5 feet), then you have to move the cardboard about 0.75 meters (2.5 feet) to see the colors of the bow. You may wonder how I arrived at these numbers. The reason is simple, the radius of a glassbow is about 28 degrees.

Once you see the colors, you can move the cardboard in a circle around the imaginary line to search for other parts of the bow. By so doing, you are mapping out the entire circular bow in portions, just as you did with the garden hose.

If you want to see the entire bow around your shadow all at once, you'll need a bigger piece of black cardboard—one full square meter will do—and with a lot more glass beads glued to it. Place the shadow of your head near the center of the cardboard. If the distance from the cardboard to your head is about 80 centimeters (about 2.5 feet), you will immediately see the entire glass bow. If you bring the cardboard too far out, e.g., 1.2 meters (4.0 feet), you will not be able to see the entire bow. The choice is yours; have fun!

If it's not a sunny day, you can try the experiment indoors, as I did in lectures, by aiming a *very* strong light—like a spotlight—at a wall, on which you've taped or hung the cardboard. Position yourself so the light is behind you, and the shadow of your head is in the center of the one square meter cardboard. If you stand 80 centimeters away from the board, you should be able to see the entire bow circling your shadow. Welcome to the glass bow!

We don't need to understand why a rainbow or fogbow or glassbow is formed in order to appreciate its beauty, of course, but understanding the physics of rainbows does give us a new set of eyes (I call this the beauty of knowledge). We become more alert to the little wonders we might just be able to spot on a foggy morning, or in the shower, or when walking by a fountain, or peeking out of an airplane window when everyone else is watching movies. I hope you will find yourself turning your back to the Sun the next time you feel a rainbow coming on, looking about 42 degrees away from the imaginary line and spotting the red upper rim of a glorious rainbow across the sky.

Here's my prediction. The next time you see a rainbow, you'll check to make sure that red is on the outside, blue is on the inside; you'll try to find the secondary bow and will confirm that the colors are reversed; you'll see that the sky is bright inside the primary bow and much darker outside of it; and if you carry a linear polarizer on you (as you always should), you will confirm that both bows are strongly polarized. You won't be able to resist it. It's a disease that will haunt you for the rest of your life. It's my fault, but I will not be able to cure you, and I'm not even sorry for that, not at all!

CHAPTER 6

The Harmonies of Strings and Winds

I took violin lessons as a ten-year-old, but I was a disaster and stopped after a year. Then in my twenties I took piano lessons, and I was a disaster again. I still cannot understand how people can read notes and convert them into music using ten fingers on different hands. I do appreciate music a lot, however, and in addition to having an emotional connection with it, I have come to understand it through physics. In fact, I love the physics of music, which starts, of course, with the physics of sound.

You probably know that sound begins with one or more very rapid vibrations of an object, like a drum surface or a tuning fork or a violin string. These vibrations are pretty obvious, right? What is actually happening when these things vibrate, however, is not so obvious, because it is usually invisible.

The back and forth motion of a tuning fork first compresses the air that is closest to it. Then, when it moves the other way, it decompresses the nearby air. This alternate pushing and pulling creates a wave in the air, a pressure wave, which we call a sound wave. This wave reaches our ears very quickly, at what we call the speed of sound: about 340 meters per second (about a mile in five seconds, or a kilometer in three). This is the speed of sound in air at room temperature. It can differ a great deal, depending on the medium it's traveling through. The speed of sound is four times faster in water and fifteen times faster in iron than in air.

The speed of light (and all electromagnetic radiation) in vacuum is a famous constant, known as c, about 300,000 kilometers per second (you may have learned it as 186,000 miles per second), but the speed of visible light is about a third slower in water.

Now to get back to the tuning fork. When the wave it produces hits our ears, it pushes our eardrums in and out at exactly the same rate of oscillations as the tuning fork presses on the air. Then, through an almost absurdly complicated process, the eardrum vibrates the bones of the middle ear, known wonderfully as the hammer, anvil, and stirrup, and they in turn produce waves in the fluid in the inner ear. These waves are then converted into electric nerve impulses sent to the brain, and your brain interprets these signals as sound. Quite a process.

Sound waves—in fact all waves—have three fundamental characteristics: frequency, wavelength, and amplitude. Frequency is the number of waves passing a given point in a given period of time. If you are watching waves in the ocean from a boat or a cruise ship, you may notice that, say, ten waves go by in a minute, so we might say they have a frequency of ten per minute. But we actually often measure frequency in oscillations per second, also known as hertz, or Hz; 200 oscillations per second is 200 hertz.

As for wavelength, this is the distance between two wave crests—or also between two wave valleys. One of the fundamental characteristics of waves is that the greater the frequency of a wave, the shorter its wavelength is; and the longer the wavelength, the lower its frequency. Here we've reached a terrifically important set of relationships in physics, those between the speed, frequency, and wavelength of waves. The wavelength of a wave is its speed divided by its frequency. This holds for electromagnetic waves (X-rays, visible light, infrared, and radio waves) as well as sound waves in a bathtub and waves in the ocean. As an example, the wavelength in air of a 440 hertz tone (middle A on the piano) is 340 divided by 440, which is 0.77 meters (about 30 inches).

If you think about this for a minute, you'll see that it makes perfect sense. Since the speed of sound is constant in any given medium (except in gases, where it depends on temperature), the more sound waves there are in a given period of time, the shorter the waves have to be to fit into that time. And the reverse is clearly also true: the fewer the waves in the same time the longer each of them has to be. For wavelength, we have different measurements for different kinds. For example, while we measure wavelengths of sound in meters, we measure the wavelengths of light in nanometers (one nanometer is a billionth of a meter).

Now what about amplitude? Think again about watching the waves out in the ocean from a boat. You will see that some waves are higher than others, even though they may have the same wavelength. This characteristic of the wave is known as its amplitude. The amplitude of a sound wave determines how loud or soft the sound will be: the greater its amplitude, the louder it is, and vice versa. This is because the larger the amplitude, the more energy a wave is carrying. As any surfer can tell you, the taller an ocean wave, the more energy it packs. When you strum guitar strings more vigorously, you are imparting more energy to them and you produce a louder sound. We measure the amplitude of water waves in meters and centimeters. The amplitude of a sound wave in air would be the distance over which the air molecules move back and forth in the pressure wave, but we never express it that way. Instead, we measure the *intensity* of sound, which is expressed in decibels. The decibel scale turns out to be quite complicated; fortunately, you don't need to know about it.

The pitch of a sound, meaning how high or low it is on the musical scale, is, on the other hand, determined by the frequency. The higher the frequency, the higher its pitch; the lower the frequency, the lower its pitch. In making music, we change the frequency (thus the pitch) all the time.

The human ear can hear a tremendous range of frequencies, from about 20 hertz (the lowest note on a piano is 27.5 hertz) all the way up to about 20,000 hertz. I have a great demonstration in my classes, in which I run a sound-producing machine, an audiometer, which can broadcast different frequencies and at different intensities. I ask students to hold their hands up as long as they can hear the sound. I gradually increase the frequency. When we get older, most of us lose our ability to hear high frequencies. My own high-frequency cutoff is near 4,000 hertz, four octaves above middle C, at the very end of the piano keyboard. But long after I'm hearing nothing, my students can hear much higher notes. I move the dial upward and still upward, to 10,000 and 15,000 hertz, and some hands start to drop. At 20,000 hertz, only about half of the hands are still up. Then I go more slowly: 21,000, 22,000, 23,000. By the time I get to 24,000 hertz, there are usually just a few hands still raised. At that point, I play a little joke on them; I turn the machine off but pretend to be raising the frequency even higher, to 27,000 hertz. One or two brave souls claim to be hearing these super high notes—until I gently puncture that balloon. It's all in good fun.

Now think about how a tuning fork works. If you hit a tuning fork harder, the number of vibrations per second of its prongs remains the same—so the frequency of the sound waves it produces stays the same. This is why it always plays the same note. However, the amplitude of the oscillation of its prongs does increase when you hit it harder. You could see this if you were to film the tuning fork as you hit it and then replay the film in slow motion. You would see the prongs of the fork move back and forth, and they would move farther the harder you hit them. Since the amplitude is increased, the note produced will be louder, but since the prongs continue to oscillate at the same frequency, the note stays the same. Isn't that weird? If you think about it for a bit, you'll see that it's exactly like the pendulum (chapter 3), where the period (the time to complete one oscillation) is independent of the amplitude of its swings.

Sound Waves in Space?

Do these relationships of sound hold true beyond Earth? Have you ever heard that there is no sound in space? This would mean that no matter how energetically you play a piano on the surface of the Moon, it wouldn't produce any sound. Can this be right? Yes, the Moon has no atmosphere; it is basically a vacuum. So you might conclude, perhaps sadly, that yes, even the most spectacular explosions of stars or galaxies colliding with each other occur in utter silence. We might even suppose that the big bang itself, the primordial explosion that created our universe nearly 14 billion years ago, took place entirely in silence. But hold on a minute. Space, like much of life, is considerably messier and more complicated than we thought even a few decades ago.

Even though any of us would quickly perish from a lack of oxygen if we tried to breathe in space, the truth is that outer space, even deep space, is not a perfect vacuum. Such terms are all relative. Interstellar and intergalactic space are millions of times closer to a vacuum than the best vacuum we can make on Earth. Still, the fact is that the matter that does float around in space has important and identifiable characteristics.

Most of this matter is called plasma: ionized gases—gases partly or completely made up of charged particles, such as hydrogen nuclei (protons) and electrons—of widely varying density. Plasma is present in our solar system, where we usually call it the solar wind, streaming outward from the Sun (the phenomenon Bruno Rossi did so much to advance our knowledge of). Plasmas are also found in stars, as well as between stars in galaxies (where we call it the interstellar medium), and even between galaxies (where we call it the intergalactic medium). Most astrophysicists believe that more than 99.9 percent of all observable matter in the universe is plasma.

Think about it. Wherever matter exists, pressure waves (thus, sound) can be produced and they will propagate. And because there are plasmas everywhere in space (also in our solar system), there are lots of sound waves out there, even though we can't possibly hear them. Our poor ears can hear a pretty wide range of frequencies—more than three orders of magnitude, in fact—but we aren't outfitted to hear the music of the heavenly spheres.

Let me give you one example. Back in 2003 physicists discovered ripples in the superhot gas (plasma) surrounding a supermassive black hole at the center of a galaxy in the Perseus cluster of galaxies, a large group of thousands of galaxies about 250 million light-years from Earth. These ripples clearly indicated sound waves, caused by the release of large amounts of energy when matter was swallowed up by the black hole. (I'll get into black holes in more detail in chapter 12.) Physicists calculated the frequency of the waves and came up with a pitch of B flat, but a B flat so low that it is 57 octaves (about a factor of 10^{17}) below middle C, whose frequency is about 262 hertz! You can see these cosmic ripples at http://science.nasa.gov/science-news/science-at-nasa/2003/09sep_blackhole sounds/.

Now let's go back to the big bang. If the primordial explosion that birthed our universe created pressure waves in the earliest matter—matter that then expanded and then cooled, creating galaxies, stars, and eventually planets—then we ought to be able to see the remnants of those sound waves. Well, physicists have calculated how far apart the ripples in the early plasma should have been (about 500,000 light-years) and how far apart they should be now, after the universe has been expanding for more than 13 billion years. The distance they came up with is about 500 million light-years.

There are two enormous galaxy-mapping projects going on right now—the Sloan Digital Sky Survey (SDSS) in New Mexico and the Two-degree Field (2dF) Galaxy Redshift Survey in Australia. They have both looked for these ripples in the distribution of galaxies and have independently found . . . guess what? They found "that galaxies are currently slightly more likely to be 500 million light-years apart than any other distance." So the big bang produced a bass gong sound that now has a wavelength of about 500 million light-years, a frequency about fifty octaves (a factor of 10^{15}) below anything our ears can hear. The astronomer Mark Whittle has played around a good bit with what he calls big bang acoustics, and you can too, by accessing his website: www.astro .virginia.edu/~dmw8f/BBA_web/index_frames.html. On the site, you can see and hear how he has simultaneously compressed time (turning 100 million years into 10 seconds) and artificially raised the pitch of the early universe fifty octaves, so you can listen to the "music" created by the big bang.

The Wonders of Resonance

The phenomenon we call resonance makes a huge number of things possible that either could not exist at all or would be a whole lot less interesting without it: not only music, but radios, watches, trampolines, playground swings, computers, train whistles, church bells, and the MRI you may have gotten on your knee or shoulder (did you know that the "R" stands for "resonance"?).

What exactly is resonance? You can get a good feeling for this by thinking of pushing a child on a swing. You know, intuitively, that you can produce large amplitudes of the swing with very little effort. Because the swing, which is a pendulum, has a uniquely defined frequency (chapter 3), if you accurately time your pushes to be in sync with that frequency, small amounts of additional push have a large cumulative impact on how high the swing goes. You can push your child higher and higher with just light touches of only a couple of fingers.

When you do this, you are taking advantage of resonance. Resonance, in physics, is the tendency of something—whether a pendulum, a tuning fork, a violin string, a wineglass, a drum skin, a steel beam, an atom, an electron, a nucleus, or even a column of air—to vibrate more powerfully at certain frequencies than at others. These we call resonance frequencies (or natural frequencies).

A tuning fork, for instance, is constructed to always vibrate at its resonance frequency. If it does so at 440 hertz, then it makes the note known as concert A, the A above middle C on the piano. Pretty much no matter how you get it vibrating, its prongs will oscillate, or move back and forth, 440 times per second.

All materials have resonance frequencies, and if you can add energy to a system or an object it may start to vibrate at these frequencies, where it takes relatively little energy input to have a very significant result. When you tap a delicate empty wineglass gently with a spoon, for example, or rub the rim with a wet finger, it will ring with a particular tone—that is a resonance frequency. Resonance is not a free lunch, though at times it looks like one. But at resonance frequencies, objects make the most efficient use of the energy you put into them.

A jump rope works on the same principle. If you've ever held one end, you know that it takes a while to get the rope swinging in a nice arc—and while you may have circled your hand around with the end to get that arc, the key part of that motion is that you are going up and down or back and forth, oscillating the rope. At a certain point, the rope starts swinging around happily in a beautiful arc; you barely have to move your hand to keep it going, and your friends can start jumping in the middle of the arc, intuitively timing their jumps to the resonant frequency of the rope.

You may not have known this on the playground, but only one person has to move her hand—the other one can simply hold on to the other end, and it works just fine. The key is that between the two of you, you've reached the rope's lowest resonance frequency, also called the fundamental. If it weren't for this, the game we know as double-dutch, in which two people swing two ropes in opposite directions, would be just about impossible. What makes it possible for two ropes to be going in opposite directions, and be held by the same people, is that each one requires very little energy to keep it going. Since your hands are the driving force here, the jump rope becomes what we call a driven oscillator. You know, intuitively, once you reach this first harmonic for the rope, that you want to stay at that frequency, so you don't move your hand any faster.

If you did, the beautiful rotating arc would break up into rope squiggles, and the jumper would quickly get annoyed. But if you had a long enough rope, and could vibrate your end more quickly, you would find that pretty soon the rope would create two arcs, one going down while the other went up, and the midpoint of the rope would stay stationary. We call that midpoint a node. At that point two of your friends could jump, one in each arc. You may have seen this in circuses. What is going on here? You have achieved a second resonance frequency. Just about everything that can vibrate has multiple resonance frequencies, which I'll discuss more in just a minute. Your jump rope has higher resonance frequencies too, which I can show you.

I use a jump rope to show multiple resonance frequencies in my class by suspending a single rope, about ten feet long, between two vertical rods. When I move one end of the rope up and down (only an inch or so), oscillating it on a rod, using a little motor whose frequency I can change, soon it will hit its lowest resonance frequency, which we call the first harmonic (it is also called the fundamental), and make one arc like the jump rope. When I oscillate the end of the rope more rapidly, we soon see two arcs, mirror images of each other. We call this the second harmonic, and it will come when we are oscillating the string at twice the rate of the first harmonic. So if the first harmonic is at 2 hertz, two vibrations per second, the second is at 4 hertz. If we oscillate the end still faster, when we reach three times the frequency of the first harmonic, in this case 6 hertz, we'll reach the third harmonic. We see the string divide equally into thirds with two points of the string (nodes) that do not move, with the arcs alternating up and down as the end goes up and down six times per second.

Remember I said that the lowest note we can hear is about 20 hertz? That's why you don't hear any music from a playground jump rope—its frequency is way too low. But if we play with a different kind of string—one on a violin or cello, say—something else entirely happens. Take a violin. You don't want me to take it, believe me—I haven't improved in the past sixty years.

In order for you to hear one long, beautiful, mournful note on a violin, there's an enormous amount of physics that has already happened. The sound of a violin, or cello, or harp, or guitar string—of any string or rope—depends on three factors: its length, its tension, and its weight. The longer the string, the lower the tension, and the heavier the string, the lower the pitch. And, of course, the converse: the shorter the string, the higher the tension, and the lighter the string, the higher the pitch. Whenever string musicians pick up their instruments after a while, they have to adjust the tension of their strings so they will produce the right frequencies, or notes.

But here's the magic. When the violinist rubs the string with a bow, she is imparting energy to the string, which somehow picks out its own resonance frequencies (from all the vibrations possible), and—here's the even more mind-blowing part—even though we cannot see it, it *vibrates simultaneously in several different resonance frequencies* (several harmonics). It's not like a tuning fork, which can only vibrate at a single frequency.

These additional harmonics (with frequencies higher than the fundamental) are often called overtones. The interplay of the varied resonant frequencies, some stronger, some weaker—the cocktail of harmonics—is what gives a violin or cello note what is known technically as its color or timbre, but what we recognize as its distinctive sound. That's the difference between the sound made by the single frequency of the tuning fork or audiometer or emergency broadcast message on the radio and the far more complex sound of musical instruments, which vibrate at several harmonic frequencies simultaneously. The characteristic sounds of a trumpet, oboe, banjo, piano, or violin are due to the distinct cocktail of harmonic frequencies that each instrument produces. I love the image of an invisible cosmic bartender, expert in creating hundreds of different harmonic cocktails, who can serve up a banjo to this customer, a kettledrum to the next, and an erhu or a trombone to the one after that.

Those who developed the first musical instruments were ingenious in crafting another vital feature of instruments that allows us to enjoy their sound. In order to hear music, the sound waves not only have to be within the frequency range you can hear, but they also must be loud enough for you to hear them. Simply plucking a string, for instance, doesn't produce enough sound to be heard at a distance. You can impart more energy to a string (and hence to the sound waves it produces) by plucking it harder, but you still may not produce a very robust sound. Fortunately, a great many years ago, millennia at least, human beings figured out how to make string instruments loud enough to be heard across a clearing or room.

You can reproduce the precise problem our ancestors faced—and then solve it. Take a foot-long piece of string, tie one end to a doorknob or drawer handle, pull on the other end until it's tight, and then pluck it with your other hand. Not much happens, right? You can hear it, and depending on the length of the string or wire, how thick it is, and how tight you hold it, you might be able to make a recognizable note. But the sound isn't very strong, right? No one would hear it in the next room. Now, if you take a plastic cup and run the string through the cup, hold the string up at an angle away from the knob or handle (so it doesn't slide toward your hand), and pluck the string, you'll hear more sound. Why? Because the string transmits some of its energy to the cup, which now vibrates at the same frequency, only it's got a much larger surface area through which to impart its vibrations to the air. As a result, you hear louder sound.

With your cup you have demonstrated the principle of a sounding board—which is absolutely essential to all stringed instruments, from guitars and bass fiddles to violins and the piano. They're usually made of wood, and they pick up the vibrations of the strings and transmit these frequencies to the air, greatly amplifying the sound of the strings.

The sounding boards are easy to see in guitars and violins. On a grand piano, the sounding board is flat, horizontal, and located underneath the strings, which are mounted on the sounding board; it stands vertically behind the strings on an upright. On a harp, the sounding board is usually the base where the strings are attached.

In class I demonstrate the workings of a sounding board in different ways. In one demonstration I use a musical instrument my daughter Emma made in kindergarten. It's one ordinary string attached to a Kentucky Fried Chicken cardboard box. You can change the tension in the string using a piece of wood. It's really great fun; as I increase the tension the pitch goes up. The KFC box is a perfect sounding board, and my students can hear the plucking of the string from quite far away. Another one of my favorite demos is with a music box that I bought many years ago in Austria; it's no bigger than a matchbox and it has no sounding board attached to it. When you turn the crank, it makes music produced by vibrating prongs. I turn the crank in class and no one can hear it, not even I! Then I place it on my lab table and turn the crank again. All the students can now hear it, even those way in the back of my large lecture hall. It always amazes me how very effective even a very simple sounding board can be.

That doesn't mean that they're not sometimes works of real art. There is a lot of secrecy about building high-quality musical instruments, and Steinway & Sons are not likely to tell you how they build the sounding boards of their world-famous pianos! You may have heard of the famous Stradivarius family in the seventeenth and eighteenth centuries who built the most fantastic and most desirable violins. Only about 540 Stradivarius violins are known to exist; one was sold in 2006 for $3.5 million. Several physicists have done extensive research on these violins in an effort to uncover the "Stradivarius secrets" in the hope that they would be able to build cheap violins with the same magic voice. You can read about some of this research at www.sciencedaily.com/releases/2009/01/090122141228.htm.

A good deal of what makes certain combinations of notes sound more or less pleasing to us has to do with frequencies and harmonics. The best-known type of note pairing, at least in Western music, is of notes where one is exactly twice the frequency of the other. We say that these notes are separated by an octave. But there are many other pleasing combinations as well: chords, thirds, fifths, and so on.

Mathematicians and "natural philosophers" have been fascinated by the beautiful numerical relationships between different frequencies since the time of Pythagoras in ancient Greece. Historians disagree over just how much Pythagoras figured out, how much he borrowed from the Babylonians, and how much his followers discovered, but he seems to get the credit for figuring out that strings of different lengths and tensions produce different pitches in predictable and pleasing ratios. Many physicists delight in calling him the very first string theorist.

Instrument makers have made brilliant use of this knowledge. The different strings on a violin, for example, all have different weights and tensions, which enable them to produce higher and lower frequencies and harmonics even though they all have about the same length. Violinists change the length of their strings by moving their fingers up and down the violin neck. When their fingers walk toward their chins, they shorten the length of any given string, increasing the frequency (thus the pitch) of the first harmonic as well as all the higher harmonics. This can get quite complex. Some stringed instruments, like the Indian sitar, have what are called sympathetic strings, extra strings alongside or underneath the playing strings that vibrate at their own resonance frequencies when the instrument is being played.

It's difficult if not impossible to see the multiple harmonic frequencies on the strings of an instrument, but I can show them dramatically by connecting a microphone to an oscilloscope, which you have probably seen on TV, if not in person. An oscilloscope shows a vibration—or oscillation—over time, on a screen, in the form of a line going up and down, above and below a central horizontal line. Intensive care units and emergency rooms are filled with them for measuring patients' heartbeats.

I always invite my students to bring their musical instruments to class so that we can see the various cocktails of harmonics that each produces.

When I hold a tuning fork for concert A up to the microphone, the screen shows a simple sine curve of 440 hertz. The line is clean and extremely regular because, as we've seen, the tuning fork produces just one frequency. But when I invite a student who brought her violin to play the same A, the screen gets a whole lot more interesting. The fundamental is still there—you can see it on the screen as the dominant sine curve—but the curve is now much more complex due to the higher harmonics, and it's different again when a student plays his cello. Imagine what happens when a violinist plays two notes at once!

When singers start demonstrating the physics of resonance by sending air through their vocal cords ("vocal folds" would be a more descriptive term), membranes vibrate and create sound waves. I ask a student to sing too, and the oscilloscope tells the same story, as similarly complicated curves pile up on the screen.

When you play the piano, the key that you press makes a hammer hit a string—a wire—whose length, weight, and tension have been set to oscillate at a given first harmonic frequency. But somehow, just like violin strings and vocal cords, the piano strings also vibrate simultaneously at higher harmonics.

Now take a tremendous thought-leap into the subatomic world and imagine super-tiny violinlike strings, much, much smaller than an atomic nucleus, that oscillate at different frequencies and different harmonics. In other words, consider the possibility that the fundamental building blocks of matter are these tiny vibrating strings, which produce all the so-called elementary particles—such as quarks, gluons, neutrinos, electrons—by vibrating at different harmonic frequencies, and in many dimensions. If you've managed to take this step, you've just grasped the fundamental proposition of string theory, the catchall term to describe the efforts of theoretical physicists over the past forty years to come up with a single theory accounting for all elementary particles and all the forces in the universe. In a way, it's a theory of "everything."

No one has the slightest idea whether string theory will succeed, and the Nobel laureate Sheldon Glashow has wondered whether it's "a theory of physics or a philosophy." But if it's true that the most basic units of the universe are the different resonance levels of unimaginably tiny strings, then the universe, and its forces and elementary particles, may resemble a cosmic version of Mozart's wonderful, increasingly complex variations on "Twinkle, Twinkle Little Star."

All objects have resonant frequencies, from the bottle of ketchup in your refrigerator to the tallest building in the world; many are mysterious and very hard to predict. If you have a car, you've heard resonances, and they didn't make you happy. Surely you've had the experience of hearing a noise while driving, and hearing it disappear when you go faster.

On my last car the dashboard seemed to hit its fundamental frequency whenever I idled at a traffic light. If I hit the gas, speeding up the engine, even if I wasn't moving, I changed the frequency of the car's vibrations, and the noise disappeared. Sometimes I would hear a new noise for a while, which usually went away when I drove faster or slower. At different speeds, which is to say different vibrating frequencies, the car—and its thousands of parts, some of which were loose, alas—hit a resonant frequency of, say, its loose muffler or deteriorating motor mounts, and they talked to me. They all said the same thing—"Take me to the mechanic; take me to the mechanic"—which I too often ignored, only to discover later the damage that these resonances had done. When I finally took the car in, I could not reproduce the awful sounds and I felt kind of stupid.

I remember when I was a student, when we had an after-dinner speaker in my fraternity we didn't like, we would take our wineglasses and run our wet fingers around the rim, something you can do at home easily, and generate a sound. This was the fundamental frequency of our wineglasses. When we got a hundred students doing it at once, it was very annoying, to be sure (this was a fraternity, after all)—but it was also very effective, and the speakers got the message.

Everyone has heard that an opera singer singing the right note loud enough can break a wineglass. Now that you know about resonance, how could that happen? It's simple, at least in theory, right? If you took a wineglass, measured the frequency of its fundamental, and then generated a sound at that frequency, what would happen? Well, most of the time, in my experience, nothing at all. I've never seen an opera singer do it. I therefore don't use an opera singer in my class. I select a wineglass, tap on it, and measure its fundamental frequency with an oscilloscope— of course it varies from glass to glass, but for the glasses I use it's always somewhere in the range of 440 to 480 hertz. I then generate electronically a sound with the *exact* same frequency of the fundamental of the wine- glass (well *exact*, of course, is never possible, but I try to get very close). I connect it to an amplifier, and slowly crank up the volume. Why increase the volume? Because the louder the sound, the more energy in the sound wave will be beating against the glass. And the greater the amplitude of the vibrations in the wineglass, the more and more the glass will bend in and out, until it breaks (we hope).

In order to show the glass vibrating, I zoom in on it with a camera and illuminate it with a strobe light, set to a slightly different frequency than the sound. It's fantastic! You see the bowl of the wineglass beginning to vibrate; the two opposite sides first contract, then push apart, and the dis- tance they move grows and grows as I increase the volume of the speaker, and sometimes I have to tweak the frequency slightly and then—*poof!*— the glass shatters. That's always the best part for the students; they can't wait for the glass to break. (You can see this online about six minutes into lecture 27 of my Electricity and Magnetism course, 8.02, at: http://ocw .mit.edu/courses/physics/8-02-electricity-and-magnetism-spring-2002/ video-lectures/lecture-27-resonance-and-destructive-resonance/.)

I also love to show students something called Chladni plates, which demonstrate, in the oddest and most beautiful ways, the effects of reso- nance. These metal plates are about a foot across, and they can be square, rectangular, or even circular, but the best are square. They are fastened to a rod or a base at their centers. We sprinkle some fine powder on the plate and then rub a violin bow along one of the sides, the whole length of the bow. The plate will start to oscillate in one or more of its reso- nance frequencies. At the peaks and valleys of the vibrating waves on the plate, the powder will shake off and leave bare metal; it will accumulate at the nodes, where the plate does not vibrate at all. (Strings have nodal points, but two-dimensional objects, like the Chladini plate, have nodal lines.)

Depending on how and where you "play" the plate by rubbing it with the bow, you will excite different resonance frequencies and make amazing, completely unpredictable patterns on its surface. In class I use a more efficient—but far less romantic—technique and hook the plate up to a vibrator. By changing the frequency of the vibrator, we see the most remarkable patterns come and go. You can see what I mean here, on YouTube: www.youtube.com/watch?v=6wmFAwqQB0g. Just try to imagine the math behind these patterns!

In the public lectures I do for kids and families, I invite the little ones to rub the plate edges with the bow—they love making such beautiful and mysterious patterns. *That's* what I'm trying to get across about physics.

The Music of the Winds

But we've left out half the orchestra! How about a flute or oboe or trombone? After all, they don't have a string to vibrate, or a soundboard to project their sound. Even though they are ancient—I saw a photograph of a 35,000-year-old flute carved out of vulture bone in the newspaper a little while ago—wind instruments are a little more mysterious than strings, partly because their mechanism is invisible.

There are different kinds of winds, of course. Some, like flutes and recorders, are open at both ends, while clarinets and oboes and trombones are closed at one end (even though they have openings for someone to blow in). But all of them make music when an infusion of air, usually from your mouth, causes a vibration of the air column *inside* the instrument.

When you blow or force air inside a wind instrument it's like plucking a guitar string or exciting a violin string with a bow—by imparting energy to the air column, you are dumping a whole spectrum of frequencies into that air cavity, and the air column itself chooses the frequency at which it wants to resonate, depending mostly on its length. In a way that is hard to imagine, but with a result that's relatively easy to calculate, the air column inside the instrument will pick out its fundamental frequency, and some of the higher harmonics as well, and start vibrating at those frequencies. Once the air column starts vibrating, it pushes and pulls on the air, just like vibrating tuning fork prongs, sending sound waves toward the ears of the listeners.

With oboes, clarinets, and saxophones, you blow on a reed, which transfers energy to the air column and makes it resonate. For flutes and piccolos and recorders, it's the way the player blows across a hole or into a mouthpiece that creates the resonance. And for brass instruments, you have to put your lips together tightly and blow a kind of buzz into the instrument—if you haven't been trained to do it, it's all but impossible. I end up just spitting into the damn thing!

If the instrument is open at both ends, like a flute or piccolo, the air column can vibrate at its harmonics, each of which is a multiple of the fundamental frequency, as was the case with the strings. But in an instrument that's closed at one end, open at the other, like a clarinet or saxophone or trumpet, the air column will only resonate at the odd-number multiples of the fundamental: three times, five times, seven times, and so on. Why this is the case gets a bit too complicated to explain here.

What's more intuitive is that the longer the air column is, the lower the frequency and the lower the pitch of the sound produced. If the length of a tube is halved, the frequency of the first harmonic will double. That's why the piccolo plays such high notes, a bassoon plays such low ones, and why the Australian didgeridoo plays really, really low tones. This general principle also explains why the smaller saxophones, the soprano and alto saxes, play higher notes than the big, long baritone sax. It's also why a pipe organ has such a range of pipe lengths—some organs can produce sounds across nine octaves. It takes an enormous tube—64 feet long (19.5 meters long, open on both sides) to produce a fundamental of about 8.7 hertz, literally below what the human ear can hear, though you can feel the vibrations. There are just two of these enormous pipes in the world, since they aren't very practical at all. A tube ten times shorter will produce a fundamental ten times higher, thus 87 hertz. A tube a hundred times shorter will produce a fundamental of about 870 hertz.

Wind instrumentalists don't just blow into their instruments. They also close or open holes in their instruments that serve to effectively shorten or lengthen the air column, thereby raising or lowering the frequency it produces. That's why, when you play around with a child's whistle, the lower tones come when you put your fingers over all the holes, lengthening the air column. The same principle holds for brass instruments. The longer the air column, even if it has to go around in circles, the lower the pitch, which is to say, the lower the frequencies of all the harmonies. The lowest-pitched tuba, known as the B-flat or BB-flat tuba, has an 18-foot-long tube with a fundamental of about 30 hertz; additional, so-called rotary valves can lower the tone to 20 hertz; the tube of a B-flat trumpet is just 4.5 feet long. The buttons on a trumpet or tuba open or close additional tubes, changing the pitch of the resonant frequencies. The trombone is the simplest to grasp visually. Pulling the slide out increases the length of the air column, lowering its resonant frequencies.

I play "Jingle Bells" on a wooden slide trombone in my class, and the students love it—I never tell them it's the only tune I can play. In fact, I'm so challenged as a musician that no matter how many times I've given the lecture, I still have to practice beforehand. I've even made marks on the slide—notes, really—numbered 1, 2, 3, and so forth; I can't even read musical notes. But as I said before, my complete lack of musical talent hasn't stopped me from appreciating music's beauty, or from having lots of fun experimenting with it.

While I'm writing this, I'm having some fun experimenting with the air column inside a one-liter plastic seltzer bottle. It's not at all a perfect column, since the bottleneck gradually widens to the full diameter of the bottle. The physics of a bottleneck can get really complicated, as you might imagine. But the basic principle of wind instrument music—the longer the air column, the lower the resonant frequencies—still holds. You can try this easily.

Fill up an empty soda or wine bottle nearly to the top (with water!) and try blowing across the top. It takes some practice, but pretty soon you will get the air column to vibrate at its resonance frequencies. The sound will be high pitched at first, but the more you drink (you see why I suggested water), the longer the column of air becomes, and the pitch of the fundamental goes down. I also find that the longer I make the air column, the more pleasing the sound is. The lower the frequency of the first harmonic, the more likely it is that I will generate additional harmonics at higher frequencies, and the sound will have a more complex timbre.

You might be thinking that it's the bottle vibrating, just as the string did, that makes the sound, and you do in fact feel the bottle vibrating, just the way you might feel a saxophone vibrate. But again, it's the air column inside that resonates. To drive home this point, consider this puzzle. If you take two identical wineglasses, one empty and one half full, and excite the first harmonic of each by tapping each glass lightly with a spoon or by rubbing its rim with a wet finger, which frequency will be higher, and why? It's not fair of me to ask this question as I have been setting you up to give the wrong answer—sorry! But perhaps you'll work it out.

The same principle is at play with those 30-inch flexible corrugated colored plastic tubes, called whirling tubes or something similar, which you've probably seen or played with. Do you remember how they work? When you start by whirling one around your head, you first hear a low-frequency tone. Of course, you expect this to be the first harmonic, just like I did when I first played with this toy. However, somehow I have never succeeded in exciting the first harmonic. It's always the second that I hear first. As you go faster, you can excite higher and higher harmonics. Advertisements online claim you can get four tones from these tubes, but you may only get three—the fourth tone, which is the fifth harmonic, takes some really, really fast whirling. I calculated the frequencies of the first five harmonics for a tube length of 30 inches and find 223 hertz (I've never gotten this one), 446 hertz, 669 hertz, 892 hertz, and 1,115 hertz. The pitch gets pretty high pretty quickly.

Dangerous Resonance

The physics of resonance reaches far beyond classroom demonstrations. Think of the different moods that music can produce with these different instruments. Musical resonance speaks to our emotions, bringing us gaiety, anxiety, calm, awe, fear, joy, sorrow, and more. No wonder we talk of experiencing emotional resonance, which can create a relationship filled with richness and depth, and overtones of understanding and tenderness and desire. It's hardly accidental that we want to be "in tune" with someone else. And how painful when we lose that resonance, either temporarily or forever, and what had felt like harmony turns into discordant interference and emotional noise. Think of the characters George and Martha in Edward Albee's *Who's Afraid of Virginia Woolf?* They fight atrociously. When the fight is one on one, they create heat, and they remain just a show for their guests. They're much more dangerous when they join forces to play get the guest.

Resonance can become powerfully destructive in physics too. The most spectacular example of destructive resonance in recent history occurred in November 1940, when a crosswind hit the main span of the Tacoma Narrows Bridge just right. This engineering marvel (which had become known as Galloping Gertie for its oscillations up and down) started to resonate powerfully. As the crosswind increased the amplitude of the bridge oscillations, the structure began to vibrate and twist, and as the twisting grew more and more extreme, the span tore apart, crashing into the water. You can watch this spectacular collapse at www.youtube .com/watch?v=j-zczJXSxnw.

Ninety years earlier, in Angers, France, a suspension bridge over the Maine River collapsed when 478 soldiers crossed it in military formation, marching in step. Their marching excited a resonance in the bridge, which snapped some corroded cables; more than 200 soldiers died when they fell into the river below. The disaster stopped suspension bridge building in France for twenty years. In 1831, British troops marching in step across the Broughton Suspension Bridge caused the bridge deck to resonate, pull out a bolt at one end of the bridge, and collapse. No one was killed, but the British army instructed all troops crossing bridges from then on to do so by breaking their marching step.

The Millennium Bridge in London opened in 2000, and many thousands of pedestrians discovered that it wobbled a good bit (it had what engineers call lateral resonance); after just a few days authorities closed the bridge for two embarrassing years while they installed dampers to control the movement generated by pedestrian footsteps. Even the great Brooklyn Bridge in New York City frightened pedestrians who packed the bridge during a 2003 electrical blackout and felt a lateral swaying in the deck that made some of them sick.

In such situations pedestrians put more weight on a bridge than the cars that are usually crossing them, and the combined motion of their feet, even if they are not in step, can start to excite a resonance vibration—a wobble—on the bridge deck. When the bridge goes one way, they compensate by stepping the other way, magnifying the amplitude of the wobble. Even engineers admit they don't know enough about the effects crowds can have on bridges. Fortunately, they know a lot about building skyscrapers that can resist the high winds and earthquakes that threaten to generate resonance frequencies that could destroy their creations. Imagine—the same principles that produced the plaintive sound of our ancestors' 35,000-year-old flute could threaten the mighty and massive Brooklyn Bridge and the tallest buildings in the world.

CHAPTER 7

The Wonders of Electricity

This works best in the winter, when the air is very dry. Make sure you're wearing a polyester shirt or sweater, then stand in front of a mirror when it's dark and start taking your shirt or sweater off. You will have anticipated that you'll hear crackling noises, just like when you pull laundry out of the dryer (unless you use one of those unromantic dryer sheets designed to reduce all that electricity). But now you will also see the glow of dozens of teeny-weeny little sparks. I love doing this because it reminds me just how close physics is to our everyday experience, if only we know how to look for it. And, as I like to point out to my students, the truth is, this little demonstration is even more fun if you do it with a friend.

You know that whenever you walk across a rug in winter and reach for a doorknob—are you wincing?—you may get a shock, and you know that it's from static electricity. You've probably even shocked a friend by shaking her hand, or felt a shock when you've handed your overcoat to a coat checker. Frankly, it feels like static electricity is everywhere in wintertime. You can feel your hair separating when you brush it, and sometimes it stands up on its own after you take your hat off. What is it about winter, and why are so many sparks flying?

The answer to all these questions begins with the ancient Greeks, who were the first to name and make a written record of the phenomenon we've come to know as electricity. Well over two thousand years ago, the Greeks knew that if you rubbed amber—the fossilized resin that they and the Egyptians made into jewelry—on a cloth, the amber could attract pieces of dry leaves. After enough rubbing, it could even produce a jolt.

I've read stories claiming that when Greeks were bored at parties, the women would rub their amber jewelry on their clothing and touch the jewelry to frogs. The frogs would jump, of course, desperately trying to escape the crazy partiers, which apparently made for great fun among the ancients. Nothing about these stories makes any sense. First off, how many parties can you imagine where there are lots of frogs waiting around to be shocked by drunken revelers? Secondly, for reasons I'll explain in a bit, static electricity doesn't work so well during the months when you're more likely to see frogs, and when the air is humid—especially in Greece. Whatever the truth of this story, what is undeniable is that the Greek word for "amber" is *electron,* so it was really the Greeks who named electricity, along with so much else of the universe, both large and small.

The European physicists of the sixteenth and seventeenth centuries, when physics was known as natural philosophy, didn't know anything about atoms or their components, but they were terrific observers, experimenters, and inventors, and some were fantastic theorists as well. You had Tycho Brahe, Galileo Galilei, Johannes Kepler, Isaac Newton, René Descartes, Blaise Pascal, Robert Hooke and Robert Boyle, Gottfried Leibniz, Christiaan Huygens—all making discoveries, writing books, disputing one another, and turning medieval scholasticism upside down.

By the 1730s, genuine scientific study of electricity (as opposed to putting on parlor tricks) was well under way in England, France, and, of course, Philadelphia. All of these experimenters had figured out that if they rubbed a glass rod with a piece of silk it would gain a charge of some kind (let's call it A)—but if they rubbed amber or rubber in the same way it would acquire a different charge (let's call it B for now). They knew that the charges were different because when they took two glass rods that they'd rubbed with silk, both charged with A, and put them near each other, they would repel each other, by some completely invisible but nevertheless palpable force. Similar objects that had been charged with charge B also repelled each other. And yet differently charged objects, say a charged glass rod (A) and a charged rubber rod (B), would attract rather than repel each other.

Charging objects by rubbing them is a truly intriguing phenomenon, and it even has a wonderful name, the "triboelectric" effect, from the Greek word for "rubbing." It feels as though the friction between the two objects is what produces the charge, but that's not the case. It turns out that some materials greedily attract charge B, while other materials can't wait to lose it, thereby creating charge A. Rubbing works because it increases the number of contact points between substances, facilitating the transfer of charge. There is a ranked list of many materials that make up the "triboelectric series" (you can find it easily online), and the farther apart two materials are on the scale, the more easily they can charge each other.

Take plastic or hard rubber that combs are typically made of. They are pretty far away from human hair in the triboelectric series, which accounts for how easily your hair can spark and stand up when you comb it in winter—especially my hair. And think about it: not only does it spark, since by vigorously combing my hair I am charging both the comb and my hair; but since the hair all picks up the same charge, whichever it is, each charged hair repels all the other like-charged hairs, and I start to resemble a mad scientist. When you scuff your shoes on a carpet, you charge yourself with A or B, depending on the material of your shoe soles and the carpet. When you get shocked by the nearest doorknob, your hand is either receiving charge from the doorknob or shooting charge to it. It doesn't matter to you which charge you have; either way, you feel the shock!

It was Benjamin Franklin—diplomat, statesman, editor, political philosopher, inventor of bifocals, swim fins, the odometer, and the Franklin stove—who introduced the idea that all substances are penetrated with what he called "electric fluid," or "electric fire." Because it seemed to explain the experimental results of his fellow natural philosophers, this theory proved very persuasive. The Englishman Stephen Gray, for instance, had shown that electricity could be conducted over distances in metal wire, so the idea of a usually invisible fluid or fire (after all, sparks do resemble fire) made good sense.

Franklin argued that if you get too much of the fire then you're positively charged, and if you have a deficiency of it then you're negatively charged. He also introduced the convention of using positive and negative signs and decided that if you rub glass with a piece of wool or silk (producing the A charge) you give it an excess of fire, and therefore it should be called positive.

Franklin didn't know what caused electricity, but his idea of an electrical fluid was brilliant as well as useful, even if not exactly correct. He maintained that if you take the fluid and bring it from one substance to another, the one with an excess becomes positively charged and, at the same time, the one from which you take the fluid becomes negatively charged. Franklin had discovered the law of conservation of electric charge, which states that you cannot truly create or get rid of charge. If you create a certain amount of positive charge, then you automatically create the same amount of negative charge. Electric charge is a zero-sum game—as physicists would say, charge is conserved.

Franklin understood, as we do today, that like charges (positive and positive, negative and negative) repel each other, and that opposite charges (positive and negative) attract. His experiments showed him that the more fire objects had, and the closer they were to each other, the stronger the forces, whether of attraction or repulsion. He also figured out, like Gray and others around the same time, that some substances conduct the fluid or fire—we now call those substances conductors—and others do not, and are therefore called nonconductors, or insulators.

What Franklin had not figured out is what the fire really consists of. If it's not fire or fluid, what is it? And why does there seem to be so much more of it in the winter—at least where I live, in the northeastern United States, shocking us right and left?

Before we take a look inside the atom to grapple with the nature of electric fire, we need to see that electricity pervades our world far more than Franklin knew—and far more than most of us realize. It not only holds together most of what we experience on a daily basis; it also makes possible everything we see and know and do. We can only think and feel and muse and wonder because electric charges jump between uncountable millions of the roughly 100 billion cells in our brains. At the same time, we can only breathe because electric impulses generated by nerves cause different muscles of our chest to contract and relax in a complicated symphony of movements. For example, and most simply, as your diaphragm contracts and drops in your thorax, it enlarges the chest cavity, drawing air into the lungs. As it relaxes and expands upward again, it pushes air out of the lungs. None of these motions would be possible without countless tiny electric impulses constantly sending messages throughout your body, in this case telling muscles to contract and then to stop contracting while others take up the work. Back and forth, back and forth, for as long as you live.

Our eyes see because the tiny cells of our retinas, the rods and cones that pick up black-white and colors, respectively, get stimulated by what they detect and shoot off electric signals through the optic nerves to our brains. Our brains then figure out whether we're looking at a fruit stand or a skyscraper. Most of our cars run on gasoline, though hybrids use increasing amounts of electricity, but there would be no gasoline used in any engine without the electricity running from the battery through the ignition to the cylinders, where electric sparks ignite controlled explosions, thousands of them per minute. Since molecules form due to electric forces that bind atoms together, chemical reactions—such as gasoline burning—would be impossible without electricity.

Because of electricity, horses run, dogs pant, and cats stretch. Because of electricity, Saran Wrap crumples, packing tape attracts itself, and the cellophane wrapping never seems to want to come off of a box of chocolates. This list is hardly exhaustive, but there's really nothing that we can imagine existing without electricity; we could not even think without electricity.

That holds true when we turn our focus to things even smaller than the microscopic cells in our bodies. Every bit of matter on Earth consists of atoms, and to really understand electricity we have to go inside the atom and briefly look at its parts: not all of them now, because that gets incredibly complicated, but just the parts we need.

Atoms themselves are so tiny that only the most powerful and ingenious instruments—scanning tunneling microscopes, atomic force microscopes, and transmission electron microscopes—can see them. (There are some astonishing images from these instruments on the web. You can see some at this link: www.almaden.ibm.com/vis/stm/gallery.html.)

If I were to take 6.5 billion atoms, roughly the same as the number of people on Earth, and line them up in a row, touching one another, I would have a line about 2 feet long. But even smaller than every atom, about ten thousand times smaller, is its nucleus, which contains positively charged protons and neutrons. The latter, as you might imagine from their name, are electrically neutral; they have no charge at all. Protons (Greek for "first one") have about the same mass as the neutrons—the inconceivably small two-billionths of a billionth of a billionth (2×10^{-27}) of a kilogram, approximately. So no matter how many protons and neutrons a nucleus has—and some have more than two hundred—it remains a real lightweight. And tiny: just about a trillionth of a centimeter in diameter.

Most important for understanding electricity, however, is that the proton has a positive charge. There's no intrinsic reason for it to be called positive, but since Franklin, physicists have called the charge left on a glass rod after it's been rubbed with silk positive, so protons are positive.

Even more important, it turns out, is the remainder of the atom, consisting of electrons—negatively charged particles that swarm around the nucleus in a cloud, at some distance by subatomic standards. If you hold a baseball in your hand, representing an atomic nucleus, the cloud of electrons around it would range as far as *half a mile* away. Clearly, most of the atom is empty space.

The negative charge of an electron is equal in strength to the positive charge of the proton. As a result, atoms and molecules that have the same number of protons and electrons are electrically neutral. When they are not neutral, when they have either an excess or deficit of electrons, we call them ions. Plasmas, as we discussed in chapter 6, are gases partially or fully ionized. Most of the atoms and molecules we deal with on Earth are electrically neutral. In pure water at room temperature only 1 in 10 million molecules are ionized.

As a consequence of Franklin's convention, when some objects have an overabundance of electrons, we say that they are negatively charged, and when they have a deficit of electrons, we say they have a positive charge. When you rub glass with a piece of silk you "rub off" (sort of) lots of electrons, so the glass ends up with a positive charge. When you rub amber or hard rubber with the same piece of silk, they collect electrons and develop a negative charge.

In most metals large numbers of electrons have escaped their atoms altogether and are more or less freely wandering around between atoms. These electrons are particularly susceptible to an external charge, either positive or negative, and when such a charge is applied, they move toward or away from it—thus creating electric current. I have a lot more to say about current, but for the time being I'll just point out that we call these materials conductors, because they easily conduct (allow the movement of) charged particles, which in this case means electrons. (Ions can also create electric currents but not in solids, and thus not in metals.)

I love the idea of electrons just ready to play, ready to move, ready to respond to positive or negative charges. In nonconductors, there's very little action of this sort; all the electrons are well fixed to their individual atoms. But that doesn't mean we can't have some fun with nonconductors—especially your garden-variety, rubber, nonconducting balloon.

You can demonstrate everything I'm talking about here by supplying yourself with a little pack of uninflated rubber balloons (thinner ones work better, like the ones you can twist into animals). Since most of you don't have glass rods sitting around, I had hoped that a water glass or wine bottle or even a lightbulb might substitute, but despite my best efforts, they don't. So why not try a large plastic or hard rubber comb? It will also be helpful to have a piece of silk, maybe an old tie or scarf, or a Hawaiian shirt your significant other has been trying to get you to throw out. But if you don't mind getting your hair mussed—for the cause of science, who would mind?—you can make use of your own hair. And you'll need to tear up some paper into, say, a few dozen or so pieces. The number doesn't matter, but they should be small, about the size of a dime or penny.

Like all static electricity experiments, these work a lot better in winter (or in afternoon desert air), when the air is dry rather than moist. Why? Because air itself is not a conductor—in fact, it's a pretty good insulator. Water, on the other hand, is a reasonable conductor, so when there's a lot of it in the air, air becomes a better conductor as well. Instead of allowing charge to build up on a rod or cloth or balloon, or your hair, humid air gradually bleeds charge away. That's why you only have a problem getting shocked on doorknobs when the air is really dry.

Invisible Induction

Assemble all your materials, and get ready to experience some of the wonders of electricity. First charge up your comb by rubbing it hard on your hair, making sure your hair is very dry, or rubbing it with the piece of silk. We know from the triboelectric series that the comb will pick up negative charge. Now, stop for a moment and think about what's going to happen as you bring the comb close to the pile of paper bits, and why. I could certainly understand if you say "nothing at all."

Then put the comb a few inches above your little mound of paper pieces. Slowly lower the comb and watch what happens. Amazing, isn't it? Try it again—it's no accident. Some of the bits of paper jump up to your comb, some stick to it for a bit and fall back down, and some stay fast. In fact, if you play around with the comb and the paper a bit, you can make the pieces of paper stand on edge, and even dance on the surface. What on earth is going on? Why do some pieces of paper stick to the comb, while others jump up, touch, and fall right back down?

These are excellent questions, with very cool answers. Here's what happens. The negative charge on the comb repels the electrons in the paper atoms so that, even though they're not free, they spend just a little more time on the far side of their atoms. When they do so, the sides of the atoms nearest the comb are just a tiny bit more positively charged than they had been before. So, the positive-leaning edge or side of the paper is attracted to the negative charge on the comb, and the very lightweight paper jumps up toward the comb. Why does their attractive force win out over the repulsive force between the comb's negative charge and the electrons in the paper? It's because the strength of electrical repulsion—and attraction—is proportional to the strength of the charges, but *inversely proportional* to the square of the distance between them. We call this Coulomb's law, named after the French physicist Charles-Augustin de Coulomb, who made this important discovery, and you will notice its astonishing similarity to Newton's law of universal gravitation. Note that we also call the basic unit of charge the coulomb, and the positive unit of charge is +1 coulomb (about 6×10^{18} protons), while the negative charge is −1 coulomb (about 6×10^{18} electrons).

Coulomb's law tells us that even a very small difference in the distance between the positive charges and the negative charges can have a large effect. Or put differently, the attractive force of the nearer charges overpowers the repelling force of the more distant charges.

We call this entire process *induction,* since what we are doing when we bring a charged object toward a neutral one is *inducing* charge on the near and far sides of the neutral object, creating a kind of charge polarization in the pieces of paper. You can see several versions of this little demonstration in my lecture for kids and their parents called "The Wonders of Electricity and Magnetism" on MIT World, which you can find here: http://mitworld.mit.edu/video/319.

As for why some bits of paper fall right back down while some stay stuck, this is also interesting. When a piece of paper touches the comb, some of the excess electrons on the comb move to the paper. When that happens, there still may be an attractive electric force between the comb and the piece of paper, but it may not be large enough anymore to counter the force of gravity, and thus the piece of paper will fall down. If the charge transfer is high, the electric force may even become repelling, in which case both the force of gravity and the electric force will accelerate the piece of paper downward.

Now blow up a balloon, knot the end so it stays blown up, and tie a string to the end. Find a place in your house where you can hang the balloon freely. From a hanging lamp, perhaps. Or you can put a weight of some kind on the string and let the balloon hang down from your kitchen table, about six inches to a foot. Charge the comb again by rubbing it vigorously with the silk or on your hair—remember, more rubbing produces a stronger charge. Very slowly, bring your comb close to the balloon. What do you think is going to happen?

Now try it. Also pretty weird, right? The balloon moves toward the comb. Just like with the paper, your comb produced some kind of separation of charge on the balloon (induction!). So what will happen when you move the comb farther away—and why? You know, intuitively, that the balloon will return to its vertical position. But now you know why, right? When the external influence disappears, the electrons no longer have any reason to hang out a little more on the far side of their respective atoms. Look what we were able to deduce just from this little bit of rubbing a comb and playing with little pieces of paper and a drugstore balloon!

Now blow up some more of the balloons. What happens when you rub one vigorously on your hair? That's right. Your hair starts to do weird things. Why? Because in the triboelectric series human hair is way at the positive end, and a rubber balloon is on the seriously negative side. In other words, rubber picks up a lot of the electrons from your hair, leaving your hair charged positively. Since like charges repel, what else can your hair do when each strand has a positive charge and wants to get away from all the other like-charged hairs? Your strands of hair are repelling one another, making them stand up. This is of course also what happens when you pull a knit hat off of your head in winter. In rubbing your hair, the hat takes lots of electrons away, leaving the strands of your hair positively charged and aching to stand up.

Back to the balloons. So you've rubbed one vigorously on your hair (rubbing it on your polyester shirt may work even better). I think you know what I'm going to suggest, right? Put the balloon against the wall, or on your friend's shirt. It sticks. Why? Here's the key. When you rub the balloon, you charge it. When you hold the balloon against the wall, which is not much of a conductor, the electrons orbiting the atoms in the wall feel the repulsive force of the balloon's negative charge and spend just a wee bit more time on the side of the atom farthest away from the balloon and a little bit less on the side closest to the balloon—that's induction!

The surface of the wall, in other words, right where the balloon is touching it, will become slightly positively charged, and the negatively charged balloon will be attracted. This is a pretty amazing result. But why don't the two charges—the positive and negative charges—just neutralize each other, with charges transferring, making the balloon immediately fall off? It's a very good question. For one thing the rubber balloon has picked up some extra electrons. They don't move around very easily in a nonconductor like rubber, so charges tend to stay put. Not only that, you're not rubbing the balloon against the wall, making lots and lots of contact. It's just sitting there, doing its attractive thing. But it's also held there by friction. Remember the Rotor carnival ride back in chapter 3? Here the electric force plays the role played by the centripetal force of the Rotor. And the balloon can stay on the wall for some time, until the charge gradually leaks off the balloon, generally onto moisture in the air. (If your balloons don't stick, the air is either too humid, making the air a better conductor, or your balloons might be too heavy—I suggested thin ones for just this reason.)

I have a ball sticking balloons on the kids who come to my public lectures. I have done this for years at kids' birthday parties, and you can have great fun with it too!

Induction works for all kinds of objects, conductors as well as insulators. You could do the comb experiment with one of those helium-filled Mylar balloons you can buy in grocery or dollar stores. As you bring the comb near the balloon, its free electrons (Mylar is a conductor) tend to move away from the negatively charged comb, leaving positively charged ions nearer the comb, which then attract the balloon toward it.

Even though we can charge rubber balloons by rubbing them on our hair or shirt, rubber is, in fact, a nearly ideal insulator—which is why it's used to coat conducting wires. The rubber keeps charge from leaking out of the wires into moist air or jumping to a nearby object—making sparks. After all, you don't want sparks jumping around in flammable environments, like the walls of your house. Rubber can and does protect us from electricity all the time. What it cannot do, however, is protect us from the most powerful form of static electricity you know: lightning. For some reason people keep repeating the myth that rubber sneakers or rubber tires can protect us from lightning. I'm not sure why these ideas still have any currency, but you're best off forgetting them *immediately!* A lightning bolt is so powerful that it doesn't care one bit about a little bit of rubber. Now you *may* be safe if lightning hits your car—probably not, in reality—but it doesn't have anything to do with the rubber tires. I'll get to that a little later.

Electric Fields and Sparks

I said before that lightning was just a big spark, a complicated spark, but still a spark. But then what, you may ask, are sparks? OK, to understand sparks we need to understand something really important about electric charge. All electric charges produce invisible electric fields, just as all masses produce invisible gravitational fields. You can sense the electric fields when you bring oppositely charged objects close to each other and you see the attraction between them. Or, when you bring like-charged objects close and see the repelling force—you are seeing the effects of the electric field between the objects.

We measure the strength of that field in units of volts per meter. Frankly, it's not easy to explain what a volt is, let alone volts per meter, but I'll give it a try. The voltage of an object is a measure of what's called its electric potential. We will assign a zero electric potential to the Earth. Thus the Earth has zero voltage. The voltage of a positively charged object is positive; it's defined as the amount of energy I have to produce to bring the positive unit of charge (+1 coulomb—which is the charge of about 6×10^{18} protons) from Earth or from any conducting object connected with the Earth (e.g., the water faucets in your house) to that object. Why do I have to produce energy to move that unit of charge? Well, recall that if that object is positively charged, it will repel the positive unit charge. Thus I have to generate energy (in physics we say I have to do work) to overcome that repelling force. The unit of energy is the joule. If I have to generate 1 joule's worth of energy, then the electric potential of that object is +1 volt. If I have to generate 1,000 joules, then the electric potential is +1,000 volts. (For the definition of 1 joule, see chapter 9.)

What if the object is negatively charged? Then its electric potential is negative and it is defined as the energy I have to produce to move the negative unit of charge (−1 coulomb—about 6×10^{18} electrons) from the Earth to that object. If that amount of energy is 150 joules, then the electric potential of the object is −150 volts.

The volt is therefore the unit of electric potential. It is named after the Italian physicist Alessandro Volta, who in 1800 developed the first electric cell, which we now call a battery. Note that a volt is *not* a unit of energy; it is a unit of energy per unit charge (joules/coulomb).

An electric current runs from a high electric potential to a lower one. How strong this current is depends on the difference in electric potential and on the electric resistance between the two objects. Insulators have a very high resistance; metals have a low resistance. The higher the voltage difference and the lower the resistance, the higher the resulting electric current. The potential difference between the two small slots in the electric wall outlets in the United States is 120 volts (it's 220 volts in Europe); however, that current is also alternating (we'll get to the matter of alternating current in the next chapter). We call the unit of current the ampere (amp), named after the French mathematician and physicist André-Marie Ampère. If a current in a wire is 1 amp, it means that everywhere through the wire a charge of 1 coulomb passes per second.

So what about sparks? How does all of this explain them? If you have scuffed your shoes a lot on the carpet, you may have built up an electric potential difference as high as about 30,000 volts between you and the Earth, or between you and the doorknob of a metal door 6 meters away from you. This is 30,000 volts over a distance of 6 meters, or 5,000 volts per meter. If you approach the doorknob, the potential difference between you and the doorknob will not change, but the distance will get smaller, thus the electric field strength will increase. Soon, as you are about to touch the doorknob, it will be 30,000 volts over a distance of about 1 centimeter. That's about 3 million volts per meter.

At this high value of the electric field (in dry air at 1 atmosphere) there will be what we call an electric breakdown. Electrons will spontaneously jump into the 1-centimeter gap, and in doing so will ionize the air. This in turn creates more electrons making the leap, resulting in an avalanche, causing a spark! The electric current shoots through the air to your finger before you can touch the doorknob. I'll bet you're cringing a bit, remembering the last time you felt such a lovely little shock. The pain you feel from a spark occurs because the electric current causes your nerves to contract, quickly and unpleasantly.

What makes the noise, the crackle, when you get a shock? That's easy. The electric current heats the air super quickly, which produces a little pressure wave, a sound wave, and that's what we hear. But sparks also produce light—even though you may not see the light during the day, though sometimes you do. How the light is produced is a little more complicated. It results when the ions created in the air recombine with electrons in the air and emit some of the available energy as light. Even if you cannot see the light from sparks (because you aren't in front of a mirror in a dark room), when you brush your hair in very dry weather you can hear the crackling noise they make.

Just think, without even trying very hard, by brushing your hair or taking off that polyester shirt, you have created, at the ends of your hair, and on the surface of your shirt, electric fields of about *3 million volts per meter*. So, if you reach for a doorknob and feel a spark at, say, 3 millimeters, then the potential difference between you and the knob was of the order of 10,000 volts.

That may sound like a lot, but most static electricity isn't dangerous at all, mainly because even with very high voltage, the current (the number of charges going through you in a given period of time) is tiny. If you don't mind little jolts, you can experiment with shocks and have some fun—and demonstrate physics at the same time. However, never stick any metal in the electric outlets in your house. That can be very very dangerous—even life threatening!

Try charging yourself up by rubbing your skin with polyester (while wearing rubber-soled shoes or flip-flops, so the charge doesn't leak to the floor). Turn off the light and then slowly move your finger closer and closer to a metal lamp or doorknob. Before they touch you ought to see a spark jump across the air between the metal and your finger. The more you charge yourself up, the greater the voltage difference you'll create between you and the doorknob, so the stronger the spark will be, and the louder the noise.

One of my students was charging himself up all the time without meaning to. He reported that he had a polyester bathrobe that he only wore in the wintertime. This turned out to be an unfortunate choice, because every time he took the robe off, he charged himself up and then got a shock when he turned off his bedside lamp. It turns out that human skin is one of the most positive materials in the triboelectric series, and polyester is one of the most negative. This is why it's best to wear a polyester shirt if you want to see the sparks flying in front of a mirror in a dark room, but not a polyester bathrobe.

To demonstrate in a rather dramatic (and very funny) way how people can get charged, I invite a student who is wearing a polyester jacket to sit on a plastic chair in front of the class (plastic is an excellent insulator). Then, while standing on a glass plate to insulate myself from the floor, I start beating the student with cat fur. Amid loud laughter of the students, the beating goes on for about half a minute. Because of the conservation of charge, the student and I will get oppositely charged, and an electric potential difference will build up between us. I show my class that I have one end of a neon flash tube in my hand. We then turn off the lights in the lecture hall, and in complete darkness I touch the student with the other end of the tube, and there is a light flash (we both feel an electric shock)! The potential difference between the student and me must have been at least 30,000 volts. The current flowing through the neon flash tube and through us discharged both of us. The demonstration is hilarious and very effective.

"Professor Beats Student" on YouTube shows the beating part of my lecture: www.organic-chemistry.com/videos-professor-beats-student -%5BP4XZ-hMHNuc%5D.cfm.

To further explore the mysteries of electric potential I use a wonderful device known as the Van de Graaff generator, which appears to be a simple metal sphere mounted on a cylindrical column. In fact, it's an ingenious device for producing enormous electric potentials. The one in my classroom generally tops out at about 300,000 volts—but they can go much higher. If you look at the first six lectures on the web in my electricity and magnetism course (8.02), you will see some of the hilarious demonstrations I can do with the Van de Graaff. You'll see me create electric field breakdown—huge sparks between the large dome of the Van de Graaff and a smaller grounded ball (thus connected with the Earth). You'll see the power of an invisible electric field to light a fluorescent tube, and you'll see that when the tube turns perpendicular to the field it turns "off." You'll even see that in complete darkness I (briefly) touch one end of the tube, making a circuit with the ground, and the light glows even more strongly. I cry out a little bit, because the shock is actually pretty substantial, even though it's not in the least bit dangerous. And if you want a real surprise (along with my students), see what happens at the end of lecture 6, as I demonstrate Napoleon's truly shocking method of testing for swamp gas. The URL is: http://ocw.mit.edu/courses/physics/8-02-electricity-and-magnetism-spring-2002/video-lectures/.

Fortunately, high voltage alone won't kill or even injure you. What counts is the current that goes through your body. Current is the amount of charge per unit of time, and as said before, we measure it in amperes. It's current that can really hurt or kill you, especially if it's continuous. Why is current dangerous? Most simply, because charges moving through your body cause your muscles to contract. At extremely low levels, electric currents make it possible for your muscles to contract, or "fire," which is vital to getting around in life. But at high levels, it causes your muscles and nerves to contract so much that they twitch uncontrollably, and painfully. At even higher levels, it causes your heart to stop beating.

It is for these reasons that one of the darker sides of the history of electricity and the human body is the use of electricity for torture—since it can cause unbearable pain—and death, of course, in the case of the electric chair. If you've seen the movie *Slumdog Millionaire*, you may remember the horrible torture scenes in the police station, in which the brutish police attach electrodes to the young Jamal, causing his body to twitch wildly.

At lower levels, current can actually be healthy. If you've ever had physical therapy for your back or shoulder, you may have had the experience of what the therapists call "electrical stimulation"—stim for short. They put conducting pads connected to an electrical power source on the affected muscle and gradually increase the current. You have the odd sensation of feeling your muscles contract and release without your doing anything at all.

Electricity is also used in more dramatic healing efforts. You've all seen the TV shows where someone uses the electric pads, known as defibrillators, to try to regularize the heartbeat of a patient in cardiac distress. At one point in my own heart surgery last year, when I went into cardiac arrest, the doctors used defibrillators to get my heart beating again—and it worked! Without defibrillators, *For the Love of Physics* would never have seen the light of day.

People disagree about the exact amount of current that's lethal, for obvious reasons: there's not too much experimenting with dangerous levels. And there's a big difference as to whether the current passes through one of your hands, for instance, or whether it goes through your brain or heart. Your hand might just burn. But pretty much everyone agrees that anything more than a tenth of an ampere, even for less than a second, can be fatal if it goes through your heart. Electric chairs used varied amounts, apparently; around 2,000 volts and from 5 to 12 amperes.

Remember when you were told as a kid not to put a fork or knife into a toaster in order to pull a piece of toast out, because you might electrocute yourself? Is that really true? Well, I just looked at the ratings of three appliances in my house: a radio (0.5 amp), my toaster (7 amps), and my espresso machine (7 amps). You can find these on a label on the bottom of most appliances. Some don't have the amperage, but you can always calculate it by dividing the wattage, the appliance's power, by the voltage, usually 120 in the United States. Most of the circuit breakers in my home are rated at between 15 and 20 amps. Whether your 120-volt appliances draw 1 or 10 amps is not really what matters. What matters is that you have to stay away from accidentally causing a short circuit and, above all, from accidentally touching with a metal object the 120 volts; if you did this shortly after you had taken a shower, it could kill you. So what does all this information add up to? Just this: when your mother told you not to put a knife into a toaster while it was plugged in, she was *right*. If you ever want to repair any of your electric appliances, make sure you unplug them first. Never forget that current can be very *dangerous*.

Divine Sparks

Of course, one of the most dangerous kinds of current is lightning, which is also one of the most remarkable of all electrical phenomena. It's powerful, not completely predictable, much misunderstood, and mysterious, all at once. In mythologies from the Greek to the Mayan, lightning bolts have been either symbols of divine beings or weapons wielded by them. And no wonder. On average, there are about 16 million thunderstorms on Earth every year, more than 43,000 every day, roughly 1,800 every hour of the day, producing about 100 lightning flashes every second, or more than *8 million* lightning flashes every day, scattered around our planet.

Lightning happens when thunderclouds become charged. Generally the top of the cloud becomes positively charged, and the bottom becomes negative. Why this is the case is not yet completely understood. There's a lot of atmospheric physics, believe it or not, that we are still learning. For now, we'll simplify and imagine a cloud with its negative charge on the side closest to the Earth. Because of induction, the ground nearest the cloud will become positively charged, generating an electrical field between the Earth and the cloud.

The physics of a lightning strike is pretty complicated, but in essence a flash of lightning (electric breakdown) occurs when the electric potential between the cloud and Earth reaches tens of millions of volts. And though we think of a bolt as shooting from a cloud down to Earth, in truth they flow *both* from the cloud to the ground and from the ground back up to the cloud. Electric currents during an average lightning bolt are about 50,000 amps (though they can be as high as a few hundred thousand amps). The maximum power during an average lightning stroke is about a trillion (10^{12}) watts. However, this lasts only for about a few tens of microseconds. The total energy released per strike is therefore rarely more than a few hundred million joules. This is equivalent to the energy that a 100-watt light bulb would consume in a month. Harvesting lightning energy is therefore not only impractical but also not too useful.

Most of us know that we can tell how far away a lightning strike is by how much time elapses between seeing the bolt and hearing the thunder. But the reason why this is true gives us a glimpse of the powerful forces at play. It has nothing to do with the explanation I heard from a student once: that the lightning makes a low pressure area of some sort, and the thunder results from air rushing into the breach and colliding with the air from the other side. In fact, it's almost exactly the reverse. The energy of the bolt heats the air to about 20,000 degrees Celsius, more than three times the surface temperature of the Sun. This superheated air then creates a powerful pressure wave that slams against the cooler air around it, making sound waves that travel through the air. Since sound waves in air travel about a mile in five seconds, by counting off the seconds you can figure out fairly easily how far away a lightning strike was.

The fact that lightning bolts heat the air so dramatically explains another phenomenon you may have experienced in lightning storms. Have you ever noticed the special smell in the air after a thunderstorm in the country, a kind of freshness, almost as if the storm had washed the air clean? It's hard to smell it in the city, because there's always so much exhaust from cars. But even if you have experienced that wonderful fragrance—and if you haven't I recommend you try to make note of it the next time you're outdoors right after a lightning storm—I'll bet you didn't know that it's the smell of ozone, an oxygen molecule made up of three oxygen atoms. Normal odorless oxygen molecules are made up of two oxygen atoms, and we call these O_2. But the terrific heat of lightning discharges blows normal oxygen molecules apart—not all of them, but

Glassbow surrounding Walter Lewin's shadow at the deCordova Museum in Massachusetts. The Astronomy Picture of the Day on September 13, 2004. *Courtesy of Walter Lewin.*

The wall of fog rising up at the BTA telescope in the Caucasus Mountains in Russia. *Courtesy of Walter Lewin.*

As the fog approached (it arrived on schedule), the sun was still up and here is the result: "Saint Walter." *Courtesy of Walter Lewin.*

A white rainbow photographed near Pike's Peak in Alaska. Notice the dark, supernumerary band on the inside. *Courtesy of Wojtek Rychlik.*

A photo of a double rainbow over the Very Large Array radio astronomy observatory in New Mexico. Notice that red is on the outside of the primary bow but it is on the inside of the secondary bow, and also how much brighter the sky is inside the primary bow than outside it. However, the sky outside the secondary is brighter than inside the secondary. The very dark area between the two bows is called Alexander's dark band. *Courtesy of Kenneth R. Lang, Tufts University, and Douglas Johnson, Battelle Observatory, Washington.*

Rainbow with repeated supernumerary bows of green and purple. *Courtesy of Andrew Dunn.*

Walter Lewin's daughter Emma valiantly assisting him on a cold winter's day to create a rainbow. *Courtesy of Walter Lewin.*

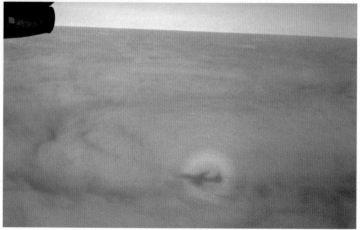

Photo of a glory surrounding the shadow of an airplane, taken by Walter Lewin. His seat (just behind the wings) is at the center of the glory. *Courtesy of Walter Lewin.*

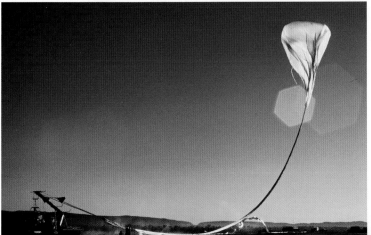

Launch of a 40-million-cubic-foot balloon from Alice Springs, Australia. *Courtesy of Walter Lewin.*

The 40-million-cubic-foot balloon at an altitude of 145,000 feet, as seen through a telescope. *Courtesy of Walter Lewin.*

Inflation of a 34-million-cubic-foot balloon shortly after sunrise in Mildura, Australia, on October 15, 1970. During this flight Lewin's group discovered GX 1+4 and its 2.3 minute periodicity. *Courtesy of Walter Lewin.*

The *Rainbow* balloon of the closing ceremonies of the 1972 Summer Olympic Games in Munich, on which Walter Lewin collaborated with Otto Piene (see chapter 15). *Courtesy of Wolf Huber.*

The Crab Nebula, with a diameter of about eleven light-years. The blue light is emitted by electrons circling around the magnetic field in the nebula. The filaments are remnants of the atmosphere of the star that exploded in 1054. Seen from the Earth it's about six times smaller than the Moon. *Courtesy of NASA and the Hubble Heritage Team; and Credit NASA/ESA/JPL/Arizona State University.*

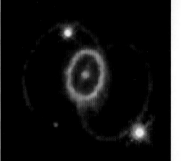

Supernova 1987A. The three rings are material expelled by the star thousands of years before the explosion. Details about the very bright inner ring are described in the text (see chapter 12). Light from the star that exploded can be seen at the center of the ring. The two white stars are not related to SN 1987A. *Courtesy of Dr. Christopher Burrows, ESA/STSci and NASA.*

An artist's impression of the Cygnus X-1 binary star system. On the left is the donor star designated HDE 226868, estimated to have about thirty times the mass of our Sun. The accreting black hole is on the right, surrounded by an accretion disk formed by the stream of gas from the donor star. The mass of the black hole is about fifteen times the mass of the Sun. *Courtesy of ESA, Hubble.*

enough to matter. And these individual oxygen atoms are unstable by themselves, so they attach themselves to normal O_2 molecules, making O_3—ozone.

While ozone smells lovely in small amounts, at higher concentrations it's less pleasant. You can often find it underneath high-voltage transmission lines. If you hear a buzzing sound from the lines, it generally means that there is some sparking, what we call corona discharge, and therefore some ozone is being created. If the air is calm, you should be able to smell it.

Now let's consider again the idea that you could survive a lightning strike by wearing sneakers. A lightning bolt of 50,000 to 100,000 amperes, capable of heating air to more than three times the surface temperature of the Sun, would almost surely burn you to a crisp, convulse you with electric shock, or explode you by converting all the water in your body instantaneously to superhot steam, sneakers or not. That's what happens to trees: the sap bursts and blows off the tree's bark. One hundred million joules of energy—the equivalent of about fifty pounds of dynamite— that's no small matter.

And what about whether you are safe inside a car when lightning strikes because of the rubber tires? You might be safe—no guarantees!— but for a very different reason. Electric current runs on the outside of a conductor, in a phenomenon called skin effect, and in a car you are effectively sitting inside a metal box, a good conductor. You might even touch the inside of your dashboard air duct and not get hurt. However, I strongly urge you not to try this; it is very dangerous as most cars nowadays have fiberglass parts, and fiberglass has no skin effect. In other words, if lightning strikes your car, you—and your car—could be in for an exceedingly unpleasant time. You might want to take a look at the short video of lightning striking a car and the photos of a van after having been hit by lightning at these sites: www.weatherimagery.com/blog/ rubber-tires-protect-lightning/ and www.prazen.com/cori/van.html. Clearly, this is not something to play around with!

Fortunately for all of us, the situation is very different with commercial airplanes. They are struck by lightning on average more than once per year, but they happily survive because of the skin effect. Watch this video at www.youtube.com/watch?v=036hpBvjoQw.

Another thing not to try in regards to lightning is the experiment so famously attributed to Benjamin Franklin: flying a kite with a key attached to it during a thunderstorm. Supposedly, Franklin wanted to test the hypothesis that thunderclouds were creating electric fire. If lightning was truly a source of electricity, he reasoned, then once his kite string got wet from the rain, it should also become a good conductor of that electricity (though he didn't use that word), which would travel down to the key tied at the base of the string. If he moved his knuckle close to the key, he should feel a spark. Now, as with Newton's claim late in life to have been inspired by an apple falling to the ground out of a tree, there is no contemporary evidence that Franklin ever performed this experiment, only an account in a letter he sent to the Royal Society in England, and another one written fifteen years later by his friend Joseph Priestley, discoverer of oxygen.

Whether or not Franklin performed the experiment—which would have been fantastically dangerous, and very likely lethal—he did publish a description of another experiment designed to bring lightning down to earth, by placing a long iron rod at the top of a tower or steeple. A few years later, the Frenchman Thomas-François Dalibard, who had met Franklin and translated his proposal into French, undertook a slightly different version of the experiment, and lived to tell the tale. He mounted a 40-foot-long iron rod pointing up into the sky, and he was able to observe sparks at the base of the rod, which was not grounded.

Professor Georg Wilhelm Richmann, an eminent scientist born in Estonia then living in St. Petersburg, Russia, a member of the St. Petersburg Academy of Sciences who had studied electrical phenomena a good deal, was evidently inspired by Dalibard's experiment, and determined to give it a try. According to Michael Brian Schiffer's fascinating book *Draw the Lightning Down: Benjamin Franklin and Electrical Technology in the Age of Enlightenment*, he attached an iron rod to the roof of his house, and ran a brass chain from the rod to an electrical measuring device in his laboratory on the first floor.

As luck—or fate—would have it, during a meeting of the Academy of Sciences in August 1753, a thunderstorm developed. Richmann rushed home, bringing along the artist who was going to illustrate Richmann's new book. While Richmann was observing his equipment, lightning struck, traveled down the rod and chain, jumped about a foot to Richmann's head, electrocuted him and threw him across the room, while also striking the artist unconscious. You can see several illustrations of the scene online, though it's not clear whether they were the creations of the artist in question.

Franklin was to invent a similar contraption, but this one was grounded; we know it today as the lightning rod. It works well to ground lightning strikes, but not for the reason Franklin surmised. He thought that a lightning rod would induce a continuous discharge between a charged cloud and a building, thus keeping the potential difference low and eliminating the danger of lightning. So confident was he in his idea that he advised King George II to put these sharp points on the royal palace and on ammunition storage depots. Franklin's opponents argued that the lightning rod would only attract lightning, and that the effect of the discharge, lowering the electric potential difference between a building and the thunderclouds, would be insignificant. The king, so the story goes, trusted Franklin and installed the lightning rods.

Not long thereafter a lightning bolt hit one of the ammunition depots, and there was very little damage. So the rod worked, but for completely the wrong reasons. Franklin's critics were right: lightning rods do attract lightning, and the discharge of the rod is indeed insignificant compared to the enormous charge on the thundercloud. But the rod really works because, if it is thick enough to handle 10,000 to 100,000 amperes, then the current will stay confined to the rod, and the charge will be transferred to the earth. Franklin was not only brilliant—he was also lucky!

Isn't it remarkable how by understanding the little crackle when we take off a sweater in winter, we can also come to some kind of understanding of the massive lightning storms that can light up the entire night sky, as well as the origin of one of the loudest, most terrifying sounds in all of nature?

In some ways we're still latter-day versions of Benjamin Franklin, trying to figure out things beyond our understanding. In the late 1980s scientists first photographed forms of lightning that occur way, way above the clouds. One kind is called red sprites and consists of reddish orange electrical discharges, 50 to 90 kilometers above the earth. And there are blue jets as well, much larger, sometimes as much as 70 kilometers long, shooting into the upper atmosphere. Since we've only known about them for a little more than twenty years, there is an awful lot we don't yet know about what causes these remarkable phenomena. Even with all we know about electricity, there are genuine mysteries on top of every thunderstorm, about 45,000 times a day.

CHAPTER 8

The Mysteries of Magnetism

F or most of us magnets are just fun, partly because they exert forces that we can feel and play with, and at the same time those forces are completely invisible. When we bring two magnets close together, they will either attract or repel each other, much as electrically charged objects do. Most of us have a sense that magnetism is deeply connected to electricity—nearly everyone interested in science knows the word *electromagnetic*, for instance—but by the same token we can't exactly explain why or how they're related. It's a huge subject, and I spend an entire introductory course on it, so we're just going to scratch the surface here. Even so, the physics of magnetism can lead us pretty quickly to some eye-popping effects and profound understandings.

Wonders of Magnetic Fields

If you take a magnet and put it in front of an older, pre-flat-screen television when it's turned on, you'll see some very cool patterns and colors across the screen. In the days before liquid crystal display (LCD) or plasma flat screens, beams of electrons shooting from the back of the TV toward the screen activated the colors, effectively painting the image on the screen. When you take a strong magnet to one of these screens, as I do in class, it will make almost psychedelic patterns. These are so compelling that even four- and five-year-olds love them. (You can easily find images of these patterns online.)

In fact, children seem to discover this on their own all the time. Anxious parents are all over the web, pleading for help in restoring their TVs after their children have run refrigerator magnets across the screens. Fortunately, most TVs come with a degaussing device that demagnetizes screens, and usually the problem goes away after a few days or a few weeks. But if it doesn't, you'll need a technician to fix the problem. So I don't recommend you put a magnet near your home TV screen (or computer monitor), unless it's an ancient TV or monitor that you don't care about. Then you might have some fun. The world-famous Korean artist Nam June Paik has created many works of art with video distortion in roughly the same way. In my class I turn on the TV and pick out a particularly awful program—commercials are great for this demonstration—and everyone loves the way the magnet completely distorts the picture.

Just as with electricity, magnetism's history goes back to ancient times. More than two thousand years ago the Greeks, the Indians, and the Chinese apparently all knew that particular rocks—which became known as lodestones—attracted small pieces of iron (just as the Greeks had found that rubbed amber would collect bits of leaves). Nowadays we call that substance magnetite, a naturally occurring magnetic mineral, in fact the most magnetic naturally occurring material on Earth. Magnetite is a combination of iron and oxygen (Fe_3O_4) and so is known as an iron oxide.

But there are lots of different kinds of magnets, not only magnetite. Iron has played such a big role in the history of magnetism, and remains such a key ingredient of many magnetically sensitive materials, that those materials that are most attracted to magnets are called ferromagnetic ("ferro" is a prefix indicating iron). These tend to be metals or metal compounds: iron itself, of course, but also cobalt, nickel, and chromium dioxide (once used heavily in magnetic tapes). Some of these can be magnetized permanently by bringing them within a magnetic field. Other materials called paramagnetic become weakly magnetic when they're placed in such a field and revert to being nonmagnetic when the field disappears. These materials include aluminum, tungsten, magnesium, and, believe it or not, oxygen. And some materials, called diamagnetic materials, develop fairly weak *opposing* magnetic fields in the presence of a magnetic field. This category includes bismuth, copper, gold, mercury, hydrogen, and table salt, as well as wood, plastics, alcohol, air, and water. (What makes certain materials ferromagnetic and some paramagnetic and others diamagnetic has to do with how the electrons are distributed around the nucleus—it's *much* too complicated to go into in detail.)

There are even liquid magnets, which are not exactly ferromagnetic liquids, but rather solutions of ferromagnetic substances that respond to magnets in very beautiful and striking ways. You can make one of these liquid magnets fairly easily; here's a link to a set of instructions: http://chemistry.about.com/od/demonstrationsexperiments/ss/liquidmagnet.htm. If you put this solution, which is fairly thick, on a piece of glass and put a magnet underneath, get ready for some remarkable results—a lot more interesting than watching iron filings line up along magnetic field lines as you may have seen in middle school.

In the eleventh century, the Chinese seem to have magnetized needles by touching them to magnetite and then suspending them from a silk thread. The needles would align themselves in the north-south direction; they aligned themselves with the magnetic field lines of the Earth. By the following century, compasses were being used for navigation both in China and as far away as the English Channel. These compasses consisted of a magnetized needle floating in a bowl of water. Ingenious, wasn't it? No matter which way the boat or ship turned, the bowl would turn but the needle would keep pointing north and south.

Nature is even more ingenious. We now know that migrating birds have tiny bits of magnetite in their bodies that they apparently use as internal compasses, helping to guide them along their migration routes. Some biologists even think that the Earth's magnetic field stimulates optical centers in some birds and other animals, like salamanders, suggesting that in some important sense, these animals can "see" the Earth's magnetic field. How cool is that?

In 1600, the remarkable physician and scientist William Gilbert—not just any doctor, but physician to Queen Elizabeth I—published his book *De Magnete, Magneticisque Corporibus, et de Magno Magnete Tellure* (*On the Magnet and Magnetic Bodies, and on That Great Magnet the Earth*), arguing that the Earth itself was a magnet, a result of his experiments with a terrella, a small magnetite sphere meant to be a model of the Earth. It was maybe a little larger than a grapefruit, and small compasses placed on it responded just as they did on the surface of the Earth. Gilbert claimed that compasses point north because the Earth was a magnet, not, as some thought, because there were magnetic islands at the North and South Poles, or that compasses were pointing toward Polaris, the North Star.

Not only was Gilbert absolutely correct that the Earth has a magnetic field, but it even has magnetic poles (just like the poles in a refrigerator magnet), which do not quite coincide with the geographic north and south poles. Not only that, but these magnetic poles wander a bit, around 15 kilometers or so every year. So in some ways the Earth does act like a simple bar magnet—an ordinary rectangular magnetized piece of metal that you can buy in a hobby shop—but in other ways it's completely different. It has taken scientists a very long time even to come up with a viable theory of why the Earth has a magnetic field. The fact that there's a lot of iron in the Earth's core isn't enough, since above a certain temperature (we call it the Curie temperature) bodies lose their ferromagnetic quality, and iron is no exception; its Curie temperature is about 770° Celsius, and we know that the core is a whole lot hotter than that!

The theory is pretty involved, and has to do with the electric currents circulating in the Earth's core and the fact that the Earth is rotating—physicists call this a dynamo effect. (Astrophysicists use the theory of these dynamo effects to explain magnetic fields in stars, including that of our own Sun, whose magnetic field *completely reverses* about every eleven years.) It may seem amazing to you, but scientists are still working on a full mathematical model of the Earth and its magnetic field; that's how complex the field is. Their work is made even thornier by the fact that there's geological evidence that the Earth's magnetic field has changed dramatically over the millennia: the poles have traveled much more than their annual stroll, and it appears that the magnetic field has also reversed itself—more than 150 times over the last 70 million years alone. Wild stuff, isn't it?

We are able to chart the Earth's magnetic field with some exactness now, thanks to satellites (such as the Danish Ørsted satellite) equipped with sensitive magnetometers. From this we know that the magnetic field reaches more than a million kilometers out into space. We also know that closer to Earth, the magnetic field produces one of the more beautiful natural phenomena in our atmosphere.

The Sun, you may remember, emits a huge stream of charged particles, mostly protons and electrons, known as the solar wind. Earth's magnetic field directs some of those particles down into our atmosphere at the magnetic poles. When these fast-moving particles, with average speeds of about 400 kilometers per second, bang into atmospheric oxygen and nitrogen molecules, some of their kinetic energy (energy of motion) gets transformed into electromagnetic energy in the form of light—oxygen releases green or red and nitrogen blue or red. You're probably guessing where I'm going—that's right: this is what produces the spectacular light show known as the aurora borealis, the northern lights, in the Northern Hemisphere and the aurora australis, or southern lights, in the Southern Hemisphere. Why do you only see these lights when you are very far north or very far south? Because the solar wind predominantly enters the Earth's atmosphere near the magnetic poles, where the magnetic field is the strongest. The reason the effects are stronger on some nights than others is that whenever there are solar eruptions, there are more particles to make the light show. When there are huge solar flares, these effects can be massive, causing what we call geomagnetic storms, producing auroras far outside the normal areas and sometimes interfering with radio transmissions, computer functioning, satellite operations, and even causing power outages.

If you don't live near the Arctic (or Antarctic) Circle, you won't see these lights very often. That's why, if you ever take an evening flight to Europe from the northeastern United States (and most flights are in the evening), you might want to try to get a seat on the left side of the plane. Since you'll be seven miles up in the atmosphere, you might see some northern lights out your window, especially if the Sun has been particularly active recently, which you can find out online. I've seen it many times in just that way, so whenever I can, I sit on the left side of the plane. I figure I can watch movies whenever I want to at home. On planes I look for the northern lights at night and glories during the day.

We are truly indebted to Earth's magnetic field, because without it, we might have suffered some serious consequences from the constant stream of charged particles bombarding our atmosphere. The solar wind might well have blasted away our atmosphere and water millions of years ago, creating conditions that would make the development of life much more difficult, if not impossible. Scientists theorize that just such a pounding by the solar wind due to Mars's weak magnetic field is what accounts for the Red Planet's thin atmosphere and comparative lack of water, an environment that human beings could inhabit only with the aid of powerful life support systems.

The Mystery of Electromagnetism

In the eighteenth century, a number of scientists began to suspect that electricity and magnetism were related in some way—even while others, such as the Englishman Thomas Young and the French scientist André-Marie Ampère, thought they had nothing to do with each other. William

Gilbert thought that electricity and magnetism were completely separate phenomena, but he nevertheless studied both simultaneously and wrote about electricity in *De Magnete* as well. He called the attractive force of rubbed amber the "electric force" (remember, the Greek word for amber was "*electron*"). And he even invented a version of the electroscope, the simplest way to measure and demonstrate the existence of static electricity. (An electroscope has a bunch of tinsel strips at the end of a metal rod. As soon as it is charged, the strips stand out away from one another: the laboratory equivalent of hat hair.)

The Bavarian Academy of Sciences invited essays on the relationship between electricity and magnetism in 1776 and 1777. People had known for some time that lightning discharges could make compasses go haywire, and none other than Benjamin Franklin himself had magnetized needles by using them to discharge Leyden jars. (Invented in the Netherlands at mid-century, the Leyden jar could store electric charges. It was an early version of the device we call a capacitor.) But while studies of electricity were exploding in the early nineteenth century, no scientist clearly linked electric current to magnetism until the Danish physicist Hans Christian Ørsted (born in 1777) made the absolutely crucial discovery bringing electricity and magnetism together. According to historian Frederick Gregory, this was probably the only time in the history of modern physics that such an enormous discovery was made in front of a class of students.

Ørsted noticed, in 1820, that an electric current flowing through a wire that was connected to a battery affected a nearby compass needle, turning it in a direction perpendicular to the wire and away from magnetic north and south. When he disconnected the wire, cutting the current flow, the needle returned to normal. It's not entirely clear whether Ørsted was conducting his experiment intentionally as part of a lecture, or whether the compass happened to be right at hand and he simply observed the astounding effect. His own accounts differ—as we've seen more than once in the history of physics.

Whether it was an accident or purposeful, this may have been the most important experiment (let's call it that) ever carried out by a physicist. He concluded reasonably that the electric current through the wire produced a magnetic field, and that the magnetic needle in the compass moved in response to that magnetic field. This magnificent discovery unleashed an explosion of research into electricity and magnetism in the nineteenth century, most notably by André-Marie Ampère, Michael Faraday, Carl Friedrich Gauss, and finally in the towering theoretical work of James Clerk Maxwell.

Since current consisted of moving electric charges, Ørsted had demonstrated that moving electric charges create a magnetic field. In 1831 Michael Faraday discovered that when he moved a magnet through a conducting coil of wire, he produced an electrical current in the coil. In effect, he showed that what Ørsted had demonstrated—that electric currents produce a magnetic field—could be turned on its head: a moving magnetic field also produces electric currents. But neither Ørsted's nor Faraday's results make any intuitive sense, right? If you move a magnet near a conducting coil—copper works great because it's so highly conductive—why on earth should you generate current in that coil? It wasn't clear at first what the importance of this discovery was. Soon afterward, the story goes, a dubious politician asked Faraday if his discovery had any practical value, and Faraday is supposed to have responded, "Sir, I do not know what it is good for. However, of one thing I am quite certain; some day you will tax it."

This simple phenomenon, which you can easily demonstrate at home, may not make any sense at all, but without exaggeration, it runs our entire economy and the entire human-made world. Without this phenomenon we would still live more or less like our ancestors in the seventeenth and eighteenth centuries. We would have candlelight, no radio, no television, no telephones, and no computers.

How do we get all this electricity that we use today? By and large we get it from power stations, which produce it with electric generators. Most fundamentally, what generators do is move copper coils through magnetic fields; we no longer move the magnets. Michael Faraday's first generator was a copper disk that he turned with a crank between the two arms of a horseshoe magnet. A brush on the outer edge of the disk ran to one wire, and a brush on the central shaft of the turning disk ran to a second wire. If he hooked the two wires up to an ammeter, it would measure the current being generated. The energy (muscle power!) he put into the system was converted by his contraption into electricity. But this generator wasn't very efficient for a variety of reasons, not the least of which was that he had to turn the copper disk with his hand. In some ways we ought to call generators energy converters. All they are doing is converting one kind of energy, in this case kinetic energy, into electric energy. There is, in other words, no free energy lunch. (I discuss the conversion of energy in more depth in the next chapter.)

Electricity into Motion

Now that we've learned how to convert motion into electricity, let's think about how to go in the other direction, converting electricity into motion. At long last, car companies are spending billions of dollars developing electric cars to do just that. They are all trying to invent efficient, powerful electric motors for these cars. And what are motors? Motors are devices that convert electric energy into motion. They all rely on a seemingly simple principle that's pretty complicated in reality: if you put a conducting coil of wire (through which a current is running) in the presence of a magnetic field, then the coil will tend to rotate. How fast it rotates depends on a variety of factors: the strength of the current, the strength of the magnetic field, the shape of the coil, and the like. Physicists say that a magnetic field exerts a torque on a conducting coil. "Torque" is the term for a force that makes things rotate.

You can visualize torque easily if you've ever changed a tire. You know that one of the most difficult parts of the operation is loosening the lug nuts holding the wheel onto the axle. Because these nuts are usually very tight, and sometimes they feel frozen, you have to exert tremendous force on the tire iron that grips the nuts. The longer the handle of the tire iron, the larger the torque. If the handle is exceptionally long, you may get away with only a small effort to loosen the bolts. You exert torque in the opposite direction when you want to tighten the nuts after you've replaced the flat tire with your spare.

Sometimes, of course, no matter how hard you push or pull, you can't budge the nut. In that case you either apply some WD-40 (and you should always carry WD-40 in your trunk, for this and many other reasons) and wait a bit for it to loosen, or you can try hitting the arm of the tire iron with a hammer (something else you should always travel with!).

We don't have to go into the complexities of torque here. All you have to know is that if you run a current through a coil (you could use a battery), and you place that coil in a magnetic field, a torque will be exerted on the coil, and it will want to rotate. The higher the current and the stronger the magnetic field, the larger the torque. This is the principle behind a direct current (DC) motor, a simple version of which is quite easy to make.

What exactly is the difference between direct current and alternating current? The polarity of the plus and minus sides of a battery does not change (plus remains plus and minus remains minus). Thus if you connect a battery to a conducting wire, a current will always flow in one direction, and this is what we call direct current. At home (in the United States), however, the potential difference between the two openings of an electrical outlet alternate with a 60-hertz frequency. In the Netherlands and most of Europe the frequency is 50 hertz. If you connect a wire, say an incandescent lightbulb or a heating coil, to an outlet in your home, the current will oscillate (from one direction to the opposite direction) with a 60-hertz frequency (thus reversing 120 times per second). This is called alternating current, or AC.

Every year in my electricity and magnetism class we have a motor contest. (This contest was first done several years before me by my colleagues and friends Professors Wit Busza and Victor Weisskopf.) Each student receives an envelope with these simple materials: two meters of insulated copper wire, two paper clips, two thumbtacks, two magnets, and a small block of wood. They have to supply a 1.5-volt AA battery. They may use any tool, they may cut the wood and drill holes, but the motor must be built only of the material that is in the envelope (tape or glue is not allowed). The assignment is to build a motor that runs as fast as possible (produces the highest number of revolutions per minute, or RPMs) from these simple ingredients. The paper clips are meant to be the supports for the rotating coil, the wire is needed to make the coil, and the magnets must be placed so as to exert a torque on the coil when current from the battery goes through it.

Let's assume you want to enter the contest, and that as soon as you connect the battery to your coil it starts to rotate in a clockwise direction. So far so good. But perhaps much to your surprise, your coil doesn't keep rotating. The reason is that every half rotation, the torque exerted on your coil reverses direction. Torque reversal will oppose the clockwise rotation; your coil may even start to briefly rotate in the counterclockwise direction. Clearly, that's not what we want from a motor. We want continuous rotation in one direction only (be it clockwise or counterclockwise). This problem can be solved by reversing the direction of the current through the coil after every half rotation. In this way the torque on the coil will always be exerted in the same direction, and thus the coil will continue to rotate in that one direction.

In building their motors, my students have to cope with the inevitable problem of torque reversal, and a few students manage to build a so-called commutator, a device that reverses the current after every half rotation. But it's complicated. Luckily there is a very clever and easy solution to the problem without reversing the current. If you can make the current (thus the torque) go to zero after every half rotation, then the coil experiences no torque at all during half of each rotation, and a torque that is always in the same direction during the other half of each rotation. The net result is that the coil keeps rotating.

I give a point for every hundred rotations per minute that a student's motor produces, up to a maximum of twenty points. Students love this project, and because they are MIT students, they have come up with some amazing designs over the years. You may want to take a shot at this yourself. You can find the directions by clicking on the pdf link to my notes for lecture 11 at http://ocw.mit.edu/courses/physics/8-02 -electricity-and-magnetism-spring-2002/lecture-notes/.

Almost all students can make a motor that turns about 400 RPM fairly easily. How do they keep the coil turning in the same direction? First of all, since the wire is completely insulated, they have to scrape the insulation off one end of the wire coil so that it always makes contact with one side of the battery—of course, it does not matter which end they choose. It's the other end of the wire that's considerably trickier. Students only want the current to flow through the coil for half of its rotation—in other words, they want to break the circuit halfway through. So they scrape *half* of the insulation off of that other end of the wire. This means there's bare wire for half of the circumference of the wire. During the times that the current stops (every half rotation), the coil continues to rotate even though there is no torque on it (there isn't enough friction to stop it in half a rotation). It takes experimentation to get the scraping just right and to figure out which half of the wire should be bare—but as I said, nearly anyone can get it to 400 RPM. And that's what I did—but I could never get much higher than 400 RPM myself.

Then some students told me what my problem was. Once the coil starts turning more than a few hundred RPM, it starts to vibrate on its supports (the paper clips), breaking the circuit frequently, and therefore interrupting the torque. So the sharper students had figured out how to take two pieces of wire to hold the ends of the coil down on the paper clips at either end while still allowing it to rotate with little friction. And that little adjustment got them, believe it or not, to 4,000 RPM!

These students are so imaginative. In almost all motors, the axis of rotation of the coil is horizontal. But one student built a motor where the axis of rotation of the coil was vertical. The best one ever got up to 5,200 RPM—powered, remember by one little 1.5-volt battery! I remember the student who won. He was a freshman, and the young man said, as he stood with me after class in front of the classroom, "Oh, Professor Lewin, this is easy. I can build you a 4,000 RPM motor in about ten minutes." And he proceeded to do it, right in front of my eyes.

But you don't need to try to create one of these. There's an even simpler motor that you can make in a few minutes, with even fewer components: an alkaline battery, a small piece of copper wire, a drywall screw (or a nail), and a small disc magnet. It's called a homopolar motor. There's a step-by-step description of how to make one, and a video of one in action right here (drop me a line if yours goes faster than 5,000 RPM): www.evilmadscientist.com/article.php/HomopolarMotor.

Just as much fun as the motor contest, in a totally different way, is another demonstration I perform in class with a 1-foot-diameter electric coil and a conducting plate. An electric current going through a coil will produce a magnetic field, as you now know. An alternating electric current (AC) in a coil will produce an alternating magnetic field. (Recall that the current created by a battery is a direct current.) Since the frequency of the electricity in my lecture hall is 60 hertz of alternating current, as it is everywhere in the United States, the current in my coil reverses every $1/120$ second. If I place such a coil just above a metal plate, the changing magnetic field (I call this the external magnetic field) will penetrate the conducting plate. According to Faraday's law, this changing magnetic field will cause currents to flow in the metal plate; we call these eddy currents. The eddy currents in turn will produce their own changing magnetic fields. Thus there will be two magnetic fields: the external magnetic field and the magnetic field produced by the eddy currents.

During about half the time in the 1/60-second cycle, the two magnetic fields are in opposite directions and the coil will be repelled by the plate; during the other half the magnetic fields will be in the same direction and the coil will be attracted by the plate. For reasons that are rather subtle, and too technical to discuss here, there is a net repelling force on the coil, which is strong enough to make the coil levitate. You can see this in the video for course 8.02, lecture 19: http://videolectures.net/

mit802s02_lewin_lec19/. Look about 44 minutes and 20 seconds into the lecture.

I figured we ought to be able to harness this force to levitate a person, and I decided that I would levitate a woman in my class, just like magicians do, by creating a giant coil, having her lie on top, and levitating her. So my friends Markos Hankin and Bil Sanford (of the physics demonstration group) and I worked hard to get enough current going through our coils, but we ended up blowing the circuit breakers every time. So we called up the MIT Department of Facilities and told them what we needed—a few thousand amps of current—and they laughed. "We'd have to redesign MIT to get you that much current!" they told us. It was too bad, since a number of women had already emailed me, offering to be levitated. I had to write them all back with regrets. But that didn't stop us, as you can see by logging on to the lecture at about 47½ minutes in. I made good on my promise; the woman just turned out to be much lighter than I'd originally planned.

Electromagnetism to the Rescue

Levitating a woman makes for a pretty good—and funny—demonstration, but magnetic levitation has a host of more amazing and much more useful applications. It is the foundation of new technologies responsible for some of the coolest, fastest, least polluting transportation mechanisms in the world.

You've probably heard of high-speed maglev trains. Many people find them utterly fascinating, since they seem to combine the magic of invisible magnetic forces with the sleekest of modern aerodynamic design, all moving at extremely high speeds. You may not have known that "maglev" stands for "magnetic levitation." But you do know that when you hold magnetic poles close together, they either attract or repel each other. The wonderful insight behind maglev-trains is that if you could find a way to control that attractive or repulsive force, you ought to be able to levitate a train above tracks and then either pull or push it at high speed. For one kind of train, which works by electromagnetic suspension (known as EMS), electromagnets on the train lift it by magnetic attraction. The trains have a C-shaped arm coming down from them; the upper part of the arm is attached to the train, while the lower arm, below the track, has magnets on its upper surface that lift the train toward the rails, which are made of ferromagnetic material.

Since you don't want the train to latch on to the rails, and since the attractive force is inherently unstable, a complicated feedback system is needed to make sure the trains remain just the right distance away from the rails, which is less than an inch! A separate system of electromagnets that switch on and off in synchronized fashion provide the train's propulsion, by "pulling" the train forward.

The other main type of maglev train system, known as electro-dynamic suspension (EDS), relies on magnetic repulsion, using remarkable devices called superconductors. A superconductor is a substance that, when kept very cold, has no electric resistance. As a result, a supercooled coil made out of superconducting material takes very little electrical power to generate a very strong magnetic field. Even more amazing, a superconducting magnet can act like a magnetic trap. If a magnet is pushed close to it, the interplay between gravity and the superconductor holds the magnet at a particular distance. As a result, maglevs that use superconductors are naturally much more stable than EMS systems. If you try to push the superconductor and the magnet together or pull them apart, you'll find it quite hard to do. The two will want to stay the same distance from each other. (There's a wonderful little video that demonstrates the relationship between a magnet and a superconductor: http://www.youtube.com/watch?v=nWTSzBWEsms.)

If the train, which has magnets on the bottom, gets too close to the track, which has superconductors in it, the increased force of repulsion pushes it away. If it gets too far away, gravity pulls it back and causes the train to move toward the track. As a result, the train car levitates in equilibrium. Moving the train forward, which also uses mostly repulsive force, is simpler than in EMS systems.

Both methods have pluses and minuses, but both have effectively eliminated the problem of friction on conventional train wheels—a major component of wear and tear—while producing a far smoother, quieter, and above all *faster* ride. (They still have to cope with the problems of air drag, which increases rapidly with the speed of the train. That's why they are designed to be so aerodynamically sleek.) The Shanghai Maglev Train, which works by means of electromagnetic suspension and opened in 2004, takes about 8 minutes to travel the 19 miles from the city to the airport, at an average speed (as of 2008) of between 139 and 156 miles per hour—though it's capable of a top speed of 268 miles per hour, faster than any other high-speed railway in the world. You can see a short video of the Shanghai train here, made by its manufacturers: www.youtube.com/watch?v=weWmTldrOyo. The highest speed ever recorded on a maglev train belong to a Japanese test track, where the JR-Maglev train hit 361 miles per hour. Here's a short piece on the Japanese train: www.youtube.com/watch?v=VuSrLvCVoVk&feature=related.

There are lots of hilarious and informative YouTube videos featuring maglev technology. This one, in which a boy levitates a spinning pencil with six magnets and a little modeling clay, features a demonstration you can reproduce easily at home: www.youtube.com/watch?v=rrRG38WpkTQ&feature=related. But also have a look at this one, using a superconductor design. It shows a model train car zipping around a track—and even has a little animated explanatory section: www.youtube.com/watch?v=GHtAwQXVsuk&feature=related.

My favorite Maglev demonstration, however, is the wonderful little spinning top known as the Levitron. You can see different versions at www.levitron.com. I have an early one in my office that has delighted hundreds of visitors.

Maglev train systems have genuine environmental advantages—they use electricity relatively efficiently and don't emit greenhouse gases in exhaust. But maglev trains don't produce something for nothing. Because most maglev tracks are not compatible with existing rail lines, maglev systems require a lot of up-front capital, which has worked against them being in widespread commercial use so far. Even so, developing more efficient and cleaner mass transit systems than what we use today is absolutely essential for our future if we're not going to cook our own planet.

Maxwell's Extraordinary Achievement

Many physicists think that James Clerk Maxwell was one of the most important physicists of all time, perhaps right behind Newton and Einstein. He contributed to an incredible range of fields in physics, from an analysis of Saturn's rings, to exploring the behavior of gases, thermodynamics, and the theory of color. But his most dazzling achievement was developing the four equations describing and linking electricity and magnetism that have become known as Maxwell's equations. These four equations only appear simple; the math behind them is pretty complicated. But if you're comfortable with integrals and differential equations, please take a look at my lectures or surf around on the web to learn about them. For our purposes, here's what Maxwell did in simpler terms.

Above all, Maxwell unified the theory of electricity and magnetism by showing these two phenomena to be just one phenomenon—electromagnetism—with different manifestations. With one very important exception, the four equations are not his "laws" or inventions; they already existed in one form or another. What Maxwell did, however, was bring them together in what we call a complete field theory.

The first of these equations is Gauss's law for electricity, which explains the relationship between electric charges and the strength and distribution of the electric fields they create. The second equation, Gauss's law for magnetism, is the simplest of the four and says several things at once. It says that there are no such things as magnetic monopoles. Magnets always have a north and south pole (we call them dipoles) as opposed to electricity which allows for electric monopoles (a monopole is either a positively charged particle or a negatively charged one). If you break one of your magnets (I have many on my refrigerator) in two pieces, each piece has a north and a south pole, and if you break it into 10,000 pieces, each has a north pole and a south pole. There is *no way* that you could end up with only a magnetic north pole in one hand and only a magnetic south pole in the other hand. However, if you have an object which is electrically charged (say, positively charged) and you break it into two pieces, both pieces can be positively charged.

Then things get really interesting. The third equation is Faraday's law, which describes how changing magnetic fields produce electric fields. You can see how this equation serves as the theoretical foundation of the electric generators I talked about earlier. The last equation is Ampère's law, which Maxwell modified in important ways. Ampère's original law showed that an electric current generated a magnetic field. But by the time he was done with it, Maxwell had added a refinement, that a changing electric field creates a magnetic field.

By playing around with the four equations, Maxwell predicted the existence of electromagnetic waves traveling through empty space. What's more, he could even calculate the speed of these waves. The truly shocking result was that their speed was the same as the speed of light. In other words, he concluded, light itself had to be an electromagnetic wave!

These scientists—Ampère, Faraday, and Maxwell—knew they were on the brink of a total revolution. Researchers had been trying to understand electricity in a serious way for a century, but now these guys were constantly breaking new ground. I sometimes wonder how they managed to sleep at night.

Maxwell's equations, because of what they brought together in 1861, were really the crowning achievement of nineteenth-century physics, most certainly for all physics between Newton and Einstein. And like all profound discoveries, they pointed the way for further efforts to try to unify fundamental scientific theories.

Ever since Maxwell, physicists have spent incalculable efforts trying to develop a single unified theory of nature's four fundamental forces:

the electromagnetic, strong nuclear, weak nuclear, and gravitational forces. Albert Einstein spent the last thirty years of his life in a failed effort to combine electromagnetism and gravity in what became known as a unified field theory.

The search for unification goes on. Abdus Salam, Sheldon Glashow, and Steven Weinberg won the Nobel Prize in 1979 for unifying electromagnetism and the weak nuclear force into what's known as the electroweak force. Many physicists are trying to unify the electroweak force and the strong nuclear force into what is called a grand unified theory, or GUT, for short. Achieving that level of unification would be a staggering accomplishment, on a par with Maxwell's. And if, somehow, somewhere, a physicist ever manages to combine gravity with GUT to create what many call a theory of everything—well, that will be the holiest of Holy Grails in physics. Unification is a powerful dream.

That's why, in my Electricity and Magnetism course, when we finally see all of Maxwell's equations in their full glory and simplicity, I project them all around in the lecture hall and I celebrate this important milestone with the students by handing out flowers. If you can handle a little suspense, you will read more about this in chapter 15.

CHAPTER 9

Energy Conservation—
Plus ça change . . .

O ne of the most popular demonstrations I've done through the years involves risking my life by putting my head directly in the path of a wrecking ball—a mini version of a wrecking ball, it must be said, but one that could easily kill me, I assure you. Whereas the wrecking balls used by demolition crews might be made from a bob, or spherical weight, of about a thousand kilos, I construct mine with a 15-kilo bob—about 33 pounds. Standing at one side of the lecture hall, with my head backed up against the wall, I hold the bob in my hands, snug up to my chin. When releasing it I must be extremely careful not to give it any kind of a push, not even a tiny little bit of a shove. Any push at all and it will surely injure me—or, as I say, possibly even kill me. I ask my students not to distract me, to make no noise, and even to stop breathing for a while—if not, I say, this could be my last lecture.

I have to confess that every time I perform this demonstration, I feel an adrenaline rush as the ball comes swinging back my way; as sure as I am that the physics will save me, it is always unnerving to stand there while it comes flying up to within a whisker of my chin. Instinctively I clench my teeth. And the truth is, I always close my eyes too! What, you may ask, what possesses me to perform this demonstration? My utter confidence in one of the most important concepts in all of physics—the law of the conservation of energy.

One of the most remarkable features of our world is that one form of energy can be converted into another form, and then into another and another, and even converted back to the original. Energy can be transformed but never lost, and never gained. In fact, this transformation happens all the time. All civilizations, not only ours but even the least technologically sophisticated, depend on this process, in many variations. This is, most obviously, what eating does for us; converting the chemical energy of food, mostly stored in carbon, into a compound called adenosine triphosphate (ATP), which stores the energy our cells can use to do different kinds of work. It's what happens when we light a campfire, converting the chemical energy stored in wood or charcoal (the carbon in each combines with oxygen) into heat and carbon dioxide.

It's what drives an arrow through the air once it's been shot from a bow, converting the potential energy, built up when you pull the bow-string back into kinetic energy, propelling the arrow forward. In a gun, it's the conversion of chemical energy from the gunpowder into the kinetic energy of rapidly expanding gas that propels bullets out of the barrel. When you ride a bicycle, the energy that pushes the pedals began as the chemical energy of your breakfast or lunch, which your body converted into a different form of chemical energy (ATP). Your muscles then use that chemical energy, converting some of it into mechanical energy, in order to contract and release your muscles, enabling you to push the pedals. The chemical energy stored in your car battery is converted to electric energy when you turn the ignition key. Some electric energy goes to the cylinders, where it ignites the gasoline mixture, releasing the chemical energy released by the gasoline as it burns. That energy is then converted into heat, which increases the pressure of the gas in the cylinder, which in turn pushes the pistons. These turn the crankshaft, and the transmission sends the energy to the wheels, making them turn. Through this remarkable process the chemical energy of the gasoline is harnessed to allow us to drive.

Hybrid cars rely in part on this process in reverse. They convert some of the kinetic energy of a car—when you step on the brakes—into electric energy that is stored in a battery and can run an electric motor. In an oil-fired furnace, the chemical energy of the oil is converted into heat, which raises the temperature of water in the heating system, which a pump then forces through radiators. In neon lights, the kinetic energy of electric charges moving through a neon gas tube is converted into visible light.

The applications are seemingly limitless. In nuclear reactors, the nuclear energy that is stored in uranium or plutonium nuclei is converted into heat, which turns water into steam, which turns turbines, which create electricity. Chemical energy stored in fossil fuels—not only oil and gasoline but also coal and natural gas—is converted into heat, and, in the case of a power plant, is ultimately converted to electrical energy.

You can witness the wonders of energy conversion easily by making a battery. There are lots of different kinds of batteries, from those in your conventional or hybrid car to those powering your wireless computer mouse and cell phone. Believe it or not, but you can make a battery from a potato, a penny, a galvanized nail, and two pieces of copper wire (each about 6 inches long, with a half-inch of insulation scraped off at each end). Put the nail most of the way into the potato at one end, cut a slit at the other end for the penny, and put the penny into the slit. Hold the end of one piece of wire on the nail (or wrap it around the nail head); hold the other piece of wire on the penny or slide it into the slit so it touches the penny. Then touch the free ends of the wires to the little leads of a Christmas tree light. It should flicker a little bit. Congratulations! You can see dozens of these contraptions on YouTube—why not give it a try?

Clearly, conversions of energy are going on around us all of the time,

but some of them are more obvious than others. One of the most counterintuitive types is that of what we call gravitational potential energy.
Though we don't generally think of static objects as having energy, they
do; in some cases quite a bit of it. Because gravity is always trying to
pull objects down toward the center of the Earth, every object that you
drop from a certain height will pick up speed. In doing so, it will lose
gravitational potential energy but it will gain kinetic energy—no energy
was lost and none was created; it's a zero sum game! If an object of mass
m falls down over a vertical distance *h*, its potential energy decreases
by an amount *mgh* (*g* is the gravitational acceleration, which is about
9.8 meters per second per second), but its kinetic energy will increase by
the same amount. If you move the object upward over a vertical distance
h, its gravitational potential energy will increase by an amount *mgh*, and
you will have to produce that energy (you will have to do work).

If a book with a mass of 1 kilogram (2.2 pounds) is on a shelf 2 meters
(about 6.5 feet) above the floor, then, when it falls to the floor, its gravitational potential energy will decrease by $1 \times 9.8 \times 2 = 19.6$ joules but its
kinetic energy will be 19.6 joules when it hits the floor.

I think the name *gravitational potential energy* is an excellent name.
Think of it this way. If I pick the book up from the floor and place it on
the shelf, it takes 19.6 joules of my energy to do so. Is this energy lost?
No! Now that the book is 2 meters above the floor, it has the "potential"
of returning that energy back to me in the form of kinetic energy—
whenever I drop it on the floor, be it the next day or the next year! The
higher the book is above the floor, the more energy is "potentially" available, but, of course I have to provide that extra energy to place the book
higher.

In a similar way, it takes energy to pull the string of a bow back when
you want to shoot an arrow. That energy is stored in the bow and it is
"potentially" available, at a time of your chosing, to convert that potential energy into kinetic energy, which gives the arrow its speed.

Now, there is a simple equation I can use to show you something quite wonderful. If you bear with me for just a bit of math, you'll see why Galileo's most famous (non)experiment works. Recall that he was said to have dropped balls of different mass (thus different weight) from the Leaning Tower of Pisa to show that their rate of falling was independent of their mass. It follows from Newton's laws of motion that the kinetic energy (KE) of a moving object is proportional both to the mass of the object and to the square of its speed; the equation for that is $KE = \frac{1}{2} mv^2$. And since we know that the change in gravitational potential energy of the object is converted to kinetic energy, then we can say that mgh equals $\frac{1}{2} mv^2$, so you have the equation $mgh = \frac{1}{2}mv^2$. If you divide both sides by m, m disappears from the equation entirely, and you have $gh = \frac{1}{2}v^2$. Then to get rid of the fraction we multiply both sides of the equation by 2, to get $2gh = v^2$. This means that v, the speed, which is what Galileo was testing for, equals the square root of $2gh$.* And note that mass has completely disappeared from the equation! It is literally not a factor— the speed does not depend on the mass. To take a specific example, if we drop a rock (of any mass) from a height of 100 meters, in the absence of air drag it will hit the ground with a speed of about 45 meters per second, or about 100 miles per hour.

Imagine a rock (of any mass) falling from a few hundred thousand miles away to the Earth. With what speed would it enter the Earth's atmosphere? Unfortunately, we cannot use the above simple equation that the speed is the square root of $2gh$ because the gravitational acceleration depends strongly on the distance to Earth. At the distance of the Moon (about 240,000 miles), the gravitational acceleration due to Earth is about 3,600 times smaller than what it is close to the surface of the Earth. Without showing you the math, take my word for it, the speed would be about 25,000 miles per hour!

Perhaps you can now understand how important gravitational potential energy is in astronomy. As I will discuss in chapter 13, when matter

---

*If you want to use this equation at home, use 9.8 for g and give h in meters; v is then in meters per second. If h is 3 meters (above the floor), the object will hit the floor at about 5.4 meters per second which is about 12 miles per hour.

falls from a large distance onto a neutron star, it crashes onto the neutron star with a speed of roughly 100,000 miles per second, yes, per second! If the rock had a mass of only 1 kilogram, its kinetic energy would then be about 13 thousand trillion (13×10^{15}) joules, which is roughly the amount of energy that a large (1,000 MW) power plant produces in about half a year.

The fact that different types of energy can be converted into one another and then back again is remarkable enough, but what is even more spectacular is that there is never any net loss of energy. Never. Amazing. This is why the wrecking ball has never killed me.

When I pull the 15 kilogram ball up to my chin over a vertical distance h, I increase its gravitational potential energy by mgh. When I drop the ball, it begins to swing across the room due to the force of gravity, and mgh is converted into kinetic energy. Here, h is the vertical distance between my chin and the lowest position of the bob at the end of the string. As the ball reaches its lowest point in the swing, its kinetic energy will be mgh. As the ball completes its arc and reaches the upper limit of its swing, that kinetic energy is converted back into potential energy— which is why, at the very height of a pendulum swing, the ball stops for a moment. If there's no kinetic energy, there's no movement. But that is for just the slightest moment, because then the ball goes back down again, on its reverse swing, and potential energy is converted again into kinetic energy. The sum of kinetic energy and potential energy is called mechanical energy, and in the absence of friction (in this case air drag on the bob), the total mechanical energy does not change—it is conserved.

This means that the ball can go no higher than the exact spot from which it was released—as long as no extra energy is imparted to it anywhere along the way. Air drag is my safety cushion. A *very* small amount of the mechanical energy of the pendulum is sucked away by air drag and converted into heat. As a result, the bob stops just one-eighth of an inch from my chin, as you can see in the video of lecture 11 from course 8.01. Susan has seen me do the demonstration three times—she shivers each time. Someone once asked me if I practiced a lot, and I always answer with what is true: that I do not have to practice as I trust the conservation of energy, 100 percent.

But if I were to give the ball the slightest little push when I let it go—say I had coughed just then and that caused me to give the ball some thrust—it would swing back to a spot a little higher than where I released it from, smashing into my chin.

The conservation of energy was discovered largely due to the work of a mid-nineteenth-century English brewer's son, James Joule. So important was his work to understanding the nature of energy that the international unit by which energy is measured, the joule, was named after him. His father had sent him and his brother to study with the famous experimental scientist John Dalton. Clearly Dalton taught Joule well. After Joule inherited his father's brewery, he performed a host of innovative experiments in the brewery's basement, probing in ingenious ways into the characteristics of electricity, heat, and mechanical energy. One of his discoveries was that electric current produces heat in a conductor, which he found by putting coils of different kinds of metal with current running through them into jars of water and measuring their changes in temperature.

Joule had the fundamental insight that heat is a form of energy, which refuted what had been the widely accepted understanding of heat for many years. Heat, it was thought, was a kind of fluid, which was called caloric—from which our contemporary word *calorie* derives—and the belief at the time was that this fluid heat flowed from areas of high concentration to low, and that caloric could never be either created or destroyed. Joule made note, though, that heat was produced in many ways that suggested it was of a different nature. For example, he studied waterfalls and determined that the water at the bottom was warmer than that at the top, and he concluded that the gravitational potential energy difference between the top and bottom of the waterfall was converted into heat. He also observed that when a paddle wheel was stirring water—a very famous experiment that Joule performed—it raised the temperature of the water, and in 1881 he came up with remarkably accurate results for the conversion of the kinetic energy of the paddle wheel into heat.

In this experiment Joule connected a set of paddles in a container of water to a pulley and a string from which he suspended a weight. As the weight lowered, the string turned the shaft of the paddles, rotating them in the water container. More technically, he lowered a mass, *m*, on a string over a distance, *h*. The change in potential energy was *mgh*, which the contraption converted into the rotational (kinetic) energy of the paddle, which then heated the water. Here is an illustration of the device:

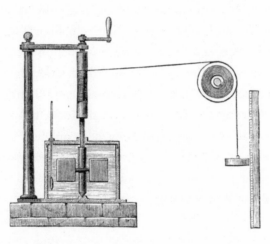

What made the experiment so brilliant is that Joule was able to calculate the exact amount of energy he was transferring to the water—which equaled *mgh*. The weight came down slowly, because the water prevented the paddle from rotating fast. Therefore the weight hit the ground with a negligible amount of kinetic energy. Thus all the available gravitational potential energy was transferred to the water.

How much is a joule? Well, if you drop a 1-kilogram object 0.1 meters (10 centimeters), the kinetic energy of that object has increased by *mgh*, which is about 1 joule. That may not sound like much, but joules can add up quite quickly. In order to throw a baseball just under 100 miles per hour, a Major League Baseball pitcher requires about 140 joules of energy, which is about the same amount of energy required to lift a bushel of 140 hundred-gram apples 1 full meter.*

One hundred forty joules of kinetic energy hitting you could be enough to kill you, as long as that energy is released quickly, and in a concentrated fashion. If it were spread out over an hour or two, you wouldn't even notice it. And if all those joules were released in a pillow hitting you hard, it wouldn't kill you. But concentrated in a bullet, say, or a rock or a baseball, in a tiny fraction of a second? A very different story.

Which brings us back to wrecking balls. Suppose you had 1,000-kilogram (1-ton) wrecking ball, which you drop over a vertical distance of 5 meters. It will convert about 50,000 joules of potential energy ($mgh = 1,000 \times 10 \times 5$) into kinetic energy. That's quite a wallop, especially if it's released in a very short time. Using the equation for kinetic energy, we can solve for speed too. At the bottom of its swing the ball would be moving at a speed of 10 meters per second (about 22 miles per hour), which is a pretty high speed for a 1-ton ball. To see this kind of energy in action, you can check out an amazing video online of a wrecking ball hitting a minivan that had strayed into a Manhattan construction zone, knocking the van over as though it were a toy car: www.lionsdenu.com/wrecking-ball-vs-dodge-mini-van/.

How Much Food Energy Do We Need?

We can come to appreciate the amazing feats of conversion of energy that keep our civilization running by considering the amount of joules involved in the most basic of our life processes. Consider, for example, that in one day a human body generates about 10 million joules of body heat. Unless you're running a fever, your body runs roughly at a temperature of 98.6 degrees Fahrenheit (37 degrees Celsius), and radiates heat in the form of infrared radiation at the rate, on average, of about 100 joules

---

*For simplicity I have used 10 meters per second for g; we do that often in physics.

per second; very roughly about 10 million joules per day. However, this does depend on air temperature and the size of the human being. The larger the person, the more energy s/he radiates per second. You can compare that to the energy radiated by a lightbulb; 1 watt is equivalent to the expenditure of 1 joule per second, so 100 joules per second equals 100 watts, which means that on average, people radiate at roughly the same level as a 100-watt lightbulb. You don't feel as hot as a lightbulb because your heat is distributed over a much larger area. When you think that an electric blanket only produces 50 watts, you now understand why, as I'm sure you already know, in winter it's much nicer to have a human being with you in bed than an electric blanket.

There are dozens of different units for energy: BTUs for air conditioners; kilowatt-hours for electricity; electron volts for atomic physics; ergs for astronomers. A BTU is about 1,055 joules; a kilowatt-hour is the equivalent of 3.6×10^6 joules; an electron volt is 1.6×10^{-19} joules; 1 erg is 10^{-7} joules. One very important unit of energy we are all familiar with is the calorie. A calorie is close to 4.2 joules. So, as our bodies generate roughly 10 million joules every day, we are expending a little over 2 million calories. But how can that be? We're supposed to eat only about 2,000 calories a day. Well, when you read *calorie* on food packages, what the label writers really mean is *kilocalorie*, a thousand calories, sometimes indicated by spelling the word *calorie* with a capital *C*. This is done for convenience, because a single calorie is a very small unit: the amount of energy required to raise the temperature of 1 gram of water 1 degree Celsius. So, in order to radiate 10 million joules per day, you have to eat roughly 2,400 kilocalories (or Calories) of food a day. And if you eat a lot more than that, well, you pay a price sooner or later. The math here is pretty unforgiving, as too many of us know but try to ignore.

What about all of the physical activity we do in a day? Don't we also have to eat to fuel that? Going up and down stairs, say, or puttering around the house, or running the vacuum cleaner? Housework can be exhausting, so we must be expending a lot of energy, right? Well, I'm afraid I have a surprise for you. It's really very disappointing. The kind of activity that you and I do in one day uses so embarrassingly little energy that you can completely neglect it if you expect to balance out food intake, unless you go to the gym for a really hard workout.

Suppose you take the stairs to climb three floors to your office instead of taking the elevator. I know plenty of people who feel virtuous for taking the stairs, but do the math. Say those three floors cover a height of about 10 meters, and you walk up them three times per day. Since I don't know you, let's give you a mass of about 70 kilograms—154 pounds. How much energy does it take to walk up those stairs three times? Let's be really virtuous—how about five times a day? Let's assume you really go out of your way. Five times a day, three floors up. The energy you would have to produce is *mgh,* where *h* is the difference in height between the first and the fourth floor. We multiply the 70 kilograms (*m*) by 10 meters per second per second (*g*) by 10 meters (*h*) by 5, since you do it five times a day, and here's what we get: 35,000 joules. Compare that to the *10 million* joules per day that your body radiates. You think you have to eat a little bit more for these lousy 35,000 joules? Forget it. It's nothing: just one-third of 1 percent of the total. But that doesn't stop marketers from making absurd claims about calorie-burning equipment. I just opened a mail-order catalog this morning that features high-end gadgets and found an ad for "wearable weights" that provide "extra calorie burning during normal daily activity." You might enjoy the feeling of your arms and legs being heavier (though I'm not sure why), and wearing them will build up muscle, but don't expect to lose significant weight by this kind of punishment!

Now a clever reader will note that of course we cannot go up the stairs five times a day without coming down. When you come down, those 35,000 joules will be released, in the form of heat in your muscles, your shoes, and the floor. If you were to jump, all of the gravitational potential energy you built up climbing the stairs would be converted to the kinetic energy of your body—and you'd probably break a bone or two. So while you had to come up with the 35,000 joules to get there, you don't get them back in a usable form when you come down, unless you can rig up a very clever device to take your kinetic energy and convert it to, say, electricity—which is exactly what hybrid cars do.

Look at it another way. Say you spread that stair climbing out over ten hours in a day, maybe once or twice in the morning, twice in the afternoon, and a final time in the early evening. In those ten hours, 36,000 seconds, you generated about 35,000 joules. This is, to be blunt, absurdly little—an average of about 1 watt. Compare that with your body, which radiates on average about 100 joules per second, or 100 watts. So, you can see, the energy burned by your stair climbing is completely negligible. It won't do anything for your waistline.

However, suppose you climb a 5,000-foot mountain instead? To do that, you would have to generate and use a million joules on top of your regular output. And a million is no longer negligible compared to 10 million. After climbing that mountain you feel legitimately hungry, and now you really do need more food. If you walk up that mountain in four hours, the average power that you have generated (power is joules per second) is substantial, an average of 70 watts during those four hours, of course. And so now the body sends an emphatic message to your brain: "I need to eat more."

You might think that since you've used 10 percent more energy over your normal 10 million joules that you would only have to eat 10 percent more (thus 240 Calories more) than you normally eat, because it's pretty obvious that a million is only 10 percent of 10 million. But that's not quite true, which you probably knew intuitively. You have to eat a good bit more than normal, because the body's food-to-energy conversion system is not particularly efficient—in physics terms. The best human beings do, on average, is 40 percent—that is, we convert at most 40 percent of our caloric intake to usable energy. The rest is lost as heat. It has to go somewhere, since energy is conserved. So to generate an extra million joules of energy to feed your mountain-climbing habit, you'll have to eat about 600 additional Calories, the rough equivalent of an extra meal per day.

Where Are We Going to Get What We Need?

The amount of energy required for our everyday life activities is astonishing to me. Suppose I wanted to take a bath, and I want to calculate how much energy it takes to heat the water. The equation is very simple; the amount of energy in kilocalories required is the mass in kilograms of the water times the temperature change in Celsius. So since a bath holds about 100 kilograms of water—that's about 26 gallons—and if we assume that the temperature increase is about 50 degrees Celsius, it takes roughly 5,000 kilocalories, or 20 million joules, of energy to produce a hot bath. Baths are lovely, but they require quite a bit of energy. The remarkable thing is that energy is still so cheap in the United States, that the bath will only cost about $1.50. Two hundred years ago, bathwater was heated with a wood fire. Firewood contains about 15 million joules per kilogram, so a family would have to get all the energy out of a kilo of wood for a single bath. While modern woodstoves can burn at 70 percent efficiency, an open fire or the stoves of 200 years ago convert wood to heat much less efficiently, and over a longer period of time, so it would probably take 5 to 10 kilos of wood to heat that 26-gallon bathtub. No wonder our ancestors bathed a lot less frequently than we do, and an entire family used the same bathwater.

Here are some figures to give you a sense of household energy usage. A space heater uses roughly 1,000 watts, which means that in the course of an hour, you expend about 3.6 million joules, or, to use the common term for measuring electricity, 1 kilowatt-hour. An electric furnace in a cold climate can use roughly 2,500 watts. A window-unit air-conditioner typically uses 1,500 watts, while a central-air system will use about 5 to 20 kilowatts. At 350 degrees Fahrenheit, an electric oven will use 2 kilowatts, while a dishwasher will use about 3.5 kilowatts. Here's an interesting comparison for you. A desktop computer with a 17-inch cathode-ray-tube monitor uses between 150 and 350 watts, while a computer and monitor in sleep mode only uses 20 watts or less. On the really low end, a clock radio uses just 4 watts. Since a 9-volt alkaline battery stores a total of about 18,000 joules, or about 5 watt-hours, one battery would power your clock-radio for a little more than an hour.

There are more than 6.5 billion people living on Earth, and we are using about 5×10^{20} joules of energy per year. Forty years after the OPEC oil embargo, 85 percent still comes from fossil fuels: coal, oil, and natural gas. The United States, with only a little more than 300 million residents, one-twentieth of the world population, is responsible for one-fifth of world energy usage. There's no way to get around this: we are energy spoilers, big energy spoilers. That's one reason I was so happy that President Obama appointed a Nobel Prize–winning physicist, Steven Chu, as his secretary of energy. If we're going to solve our energy problems, we're going to need to pay attention to the physics of energy.

For example, there is much hope being placed in the potential for solar energy, and I am all for developing it vigorously. But we must beware of the limitations we are up against. There is no question that the Sun is a wonderful source of energy. It produces 4×10^{26} watts—4×10^{26} joules per second—of power, most of it in visible light and in the infrared part of the spectrum. Since we know the distance between the Earth and the Sun (150 million kilometers), we can calculate how much of that power reaches the Earth. It's about 1.7×10^{17} watts, or about 5×10^{24} joules per year. If you point a one-square-meter panel directly at the Sun (no clouds!), that panel would receive roughly 1,200 watts (I have assumed here that about 15 percent of the incoming power is reflected and absorbed by the Earth's atmosphere). An easy number to work with is 1,000 watts (1 kilowatt) per square meter *pointed directly at the Sun* in the absence of clouds.

The potential for solar power would seem tremendous. It would take about 2×10^{10} square meters to harvest enough solar energy for the world's energy needs. That's about five times the area of my home country, Holland—not a very big country at all.

However, there is a catch. There are day and night, which we haven't allowed for yet. We just assumed that the Sun was always there. There are clouds, too. And if your solar panels are not movable, then they cannot remain pointed at the Sun all the time. Where you are situated on the Earth also matters. Countries at the equator receive more energy (they are hotter, after all) than more northern countries (in the Northern Hemisphere) or more southern ones (in the Southern Hemisphere).

Then we need to take into account the efficiency of the units with which you capture the solar energy. There are lots of different technologies, more all the time, but the maximum efficiency of practical silicon solar cells (as opposed to those made with expensive materials) is about 18 percent. If you use solar energy to directly heat water (without first converting it to electric energy), the efficiency is much higher. An oil-fired furnace, by comparison, even one that's not so new, can easily reach an efficiency of 75 to 80 percent. So if you take all those limiting factors into account, you would need an area more like a trillion square meters, roughly 400,000 square miles, an area about three times larger than Germany. And we haven't even considered the cost of building the arrays to collect and convert all that solar power to electricity. At the moment it costs about twice as much to extract electricity from the Sun as it does to extract it from fossil fuels. Not only would the cost of converting to solar power be staggering, such a project is simply beyond our present technological capability or political will. That's why solar power will play a growing but relatively small role in the world economy for some time.

On the other hand, if we start now, we could make enormous strides in the next four decades. Greenpeace International and the International Energy Agency estimated in 2009 that with very substantial government subsidies, solar power could meet "up to 7 percent of the world's power needs by 2030 and fully one-quarter by 2050." *Scientific American* magazine argued several years ago that a crash program and more than $400 billion in subsidies over the next forty years could result in solar power providing 69 percent of the United States' electricity, and 35 percent of its total energy needs.

What about wind power? After all, wind power has been used as long as humans have put sails into the wind. Windmills have been around way longer than electric power, maybe even a thousand years longer. And the principle of getting energy from nature and converting it into a different kind of energy for human use was exactly the same, whether it was in thirteenth-century China, even more ancient Iran, or twelfth-century Europe. In all of these places windmills helped do some of the hardest chores human beings took on: lifting water for drinking or crop irrigation, or grinding grains between large stones in order to make flour. It takes wind energy to power a windmill, whether or not it's making electricity.

As a producer of electricity, wind energy is readily available, utterly renewable, and produces no greenhouse gas emission. In 2009, wind energy production worldwide was 340 terawatt-hours (a terawatt-hour is one trillion watt-hours), which is about 2 percent of the world's electric consumption. And it is growing rapidly; in fact, electricity production from wind has doubled in the past three years.

What about nuclear energy? Nuclear energy is much more plentiful than we are generally aware. It is, in fact, all around us, every day. Window glass contains radioactive potassium-40, which has a half-life of 1.2 billion years, and energy produced by its decay helps to heat the Earth's core. All the helium in the atmosphere was produced by the radioactive decay of naturally occurring isotopes in the Earth. What we call alpha decay is in fact the emission of a helium nucleus from a larger unstable nucleus.

I have a very special, very large collection of Fiestaware, which is American tableware—dishes, bowls, saucers, and cups—designed and manufactured starting in the 1930s. I love to bring a few of these plates into class and show them to my students. The orange ones, in particular, which are called "Fiesta red," have uranium oxide in them, since it was a common ingredients in ceramic glazes. I hold a plate near a Geiger counter, and it begins to beep rapidly. The uranium in the plate emits gamma rays as a result of the process we call fission, which is the same process that drives nuclear reactors. After this demonstration, I always invite students to come to dinner at my home, but strangely I have never gotten any takers.

Fission, or splitting of heavy nuclei, generates large amounts of energy, whether in a nuclear reactor, in which the chain reactions splitting uranium-235 nuclei are controlled, or in an atomic bomb, in which the chain reactions are uncontrolled and produce tremendous destruction. A nuclear power plant that produces about a billion joules per second (10^9 watts, or 1,000 megawatts) consumes about 10^{27} uranium-235 nuclei in a year, which amounts to only about 400 kilograms of uranium-235.

However, only 0.7 percent of natural uranium consists of uranium 235 (99.3 percent is uranium-238). Therefore, nuclear power plants use *enriched* uranium; the degree of enrichment varies, but a typical number is 5 percent. This means that instead of 0.7 percent uranium-235, their uranium fuel rods contain 5 percent uranium-235. Thus a 1,000-megawatt nuclear reactor will consume about 8,000 kilograms uranium per year, of which about 400 kilograms is uranium-235. In comparison, a 1,000-megawatt fossil-fuel power plant will consume about 5 billion kilograms of coal per year.

The enrichment of uranium is costly; it's done with thousands of centrifuges. Weapon's-grade uranium is enriched to at least 85 percent uranium-235. Perhaps you now understand why the world is very worried about countries that enrich uranium to an unspecified degree that cannot be verified!

In nuclear power plants, the heat produced by the controlled chain reactions turns water into steam, which then drives a steam turbine, producing electricity. A nuclear power plant's efficiency converting nuclear energy into electricity is about 35 percent. If you read that a nuclear power plant produces 1,000 megawatts, you do not know whether it is 1,000 megawatts total power (of which ⅓ is converted to electrical energy and of which ⅔ is lost as heat), or whether it's all electric power in which case the total plant's power is about 3,000 megawatts. It makes a big difference! I read yesterday in the news that Iran is shortly going to put on line a nuclear power plant that will produce 1,000 megawatts of electricity (that's clear language!).

As concern about global warming has increased dramatically in the past few years, the nuclear energy option is coming back into fashion—unlike power plants burning fossil fuels, nuclear plants don't emit much in the way of greenhouse gases. There are already more than a hundred nuclear power plants in the United States, producing about 20 percent of the energy we consume. In France this number is about 75 percent. Worldwide, about 15 percent of the total electric energy consumed is produced in nuclear plants. Different countries have different policies regarding nuclear power, but building more plants will require a great deal of political persuasion due to the fear generated by the infamous nuclear accidents at Three Mile Island and Chernobyl. The plants are also *very* expensive: estimates range from $5 to $10 billion per plant in the United States, and around $2 billion in China. Finally, storing the radioactive waste from nuclear plants remains an enormous technological and political problem.

Of course, we still have massive amounts of fossil fuel on Earth, but we are using it up much, much faster than nature can create it. And the world population continues to grow, while energy-intensive development is proceeding at an extremely rapid clip in many of the largest growth countries, like China and India. So there really is no way around it. We have a very serious energy crisis. What should we do about it?

Well, one important thing is to become more aware of just how much energy we use every day, and to use less. My own energy consumption is quite modest, I think, although since I live in the United States, I'm sure I also consume four or five times more than the average person in the world. I use electricity; I heat my house and water with gas, and I cook with gas. I use my car—not very much, but I do use some gasoline. When I add that all up, I think I consumed (in 2009) on average about 100 million joules (30 kilowatt-hours) per day, of which about half was electrical energy. This is the energy equivalent of having about two hundred slaves working for me like dogs twelve hours a day. Think about that. In ancient times only the richest royalty lived like this. What luxurious, incredible times we live in. Two hundred slaves are working for me every single day, twelve hours a day without stopping, all so that I can live the way I live. For 1 kilowatt-hour of electricity, which is 3.6 million joules, I pay a mere 25 cents. So my entire energy bill (I included gas and gasoline, as their price per unit energy is not very different) for those two hundred slaves was, on average, about $225 a month; that's about $1 per slave per month! So a change of consciousness is vital. But that will only get us so far.

Changing habits to use more energy-conserving devices, such as compact fluorescent lights (CFLs) instead of incandescent lights, can make a large difference. I got to see the change I could make in quite a dramatic fashion. My electric consumption at my home in Cambridge was 8,860 kilowatt-hours in 2005 and 8,317 kilowatt-hours in 2006. This was for lighting, air-conditioning, my washing machine, and the dryer (I use gas for hot water, cooking, and heating). In mid-December of 2006, my son, Chuck (who is the founder of New Generation Energy), gave me a wonderful present. He replaced all the incandescent lightbulbs (a total of seventy-five) in my house with fluorescent bulbs. My electricity consumption dropped dramatically in 2007 to 5,251 kilowatt-hours, 5,184 kilowatt-hours in 2008, and 5,226 kilowatt-hours in 2009. This *40 percent reduction* in my electricity consumption lowered my yearly bill by about $850. Since lighting alone accounts for about 12 percent of U.S. residential electric energy use and 25 percent of commercial use, it's clearly the way to go!

Following a similar path, the Australian government started to make plans in 2007 to replace all incandescent lightbulbs in the country with fluorescent ones. This would not only substantially reduce Australia's greenhouse gas emission, but it would also reduce energy bills in every household (as it did in mine). We still need to do more, though.

I think the only way that we might survive while keeping anything like our current quality of life is by developing nuclear fusion as a reliable, serious energy source. Not fission—whereby uranium and plutonium nuclei break up into pieces and emit energy, which powers nuclear reactors—but fusion, in which hydrogen atoms merge together to create helium, releasing energy. Fusion is the process that powers stars—and thermonuclear bombs. Fusion is the most powerful energy-producing process per unit of mass we know of—except for matter and antimatter colliding (which has no potential for energy generation).

For reasons that are quite complicated, only certain types of hydrogen (deuterium and tritium) are well suited for fusion reactors. Deuterium (whose nucleus contains one neutron as well as one proton) is readily available; about one in every six thousand hydrogen atoms on Earth is deuterium. Since we have about a billion cubic kilometers of water in our oceans, the supply of deuterium is pretty much unlimited. There is no naturally occurring tritium on Earth (it's radioactive with a half life of about twelve years), but it is easily produced in nuclear reactors.

The real problem is how to create a functioning, practical, controlled fusion reactor. It's not at all clear that we will ever succeed in doing so. In order to get hydrogen nuclei to fuse, we need to create, here on Earth, temperatures in the 100-million-degree range, approximating the temperature at the core of stars.

Scientists have been working hard on fusion for many years—and I think they are working harder on it now that more and more governments seem genuinely convinced that the energy crisis is real. It's a big problem, for sure. But I'm an optimist. After all, in my professional lifetime I've seen changes in my field that have been absolutely mind-blowing, turning our notions of the universe upside down. Cosmology, for instance, which used to be mostly speculation and a little bit of science, has now become a genuine experimental science, and we know an enormous amount about the origins of our universe. In fact, we now live in what many call the golden age of cosmology.

When I began to do research in X-ray astronomy, we knew of about a dozen X-ray sources in deep space. Now we know of many tens of thousands. Fifty years ago the computing capacity in your four-pound laptop would have taken up most of the building at MIT where I have my office. Fifty years ago astronomers relied on ground-based optical and radio telescopes—that was it! Now we not only have the Hubble Space Telecope, we've had a string of X-ray satellite observatories, gamma ray observatories, and we're using and building new neutrino observatories! Fifty years ago even the likelihood of the big bang was not a settled issue. Now we not only think we know what the universe looked like in the first one-millionth of a second after the big bang—we confidently study astronomical objects more than 13 *billion* years old, objects formed in the first 500 million years after the explosion that created our universe. Against the backdrop of these immense discoveries and transformations, how can I not think scientists will solve the problem of controlled fusion? I don't want to trivialize the difficulties, or the importance of doing so soon, but I do believe it's only a question of time.

CHAPTER 10

X-rays from Outer Space!

The heavens have always provided a daily and nightly challenge to human beings seeking to understand the world around us, which is one reason physicists have always been entranced by astronomy. "What is the Sun?" we wonder. "Why does it move?" And what about the Moon, the planets, and the stars? Think about what it took for our ancestors to figure out that the planets were different from the stars; that they orbited the Sun; and that those orbits could be observed, charted, explained, and predicted. Many of the greatest scientific minds of the sixteenth and seventeenth centuries—among them Nicolaus Copernicus, Galileo Galilei, Tycho Brahe, Johannes Kepler, Isaac Newton—were compelled to turn their gaze to the heavens to unlock these nightly mysteries. Imagine how exciting it must have been for Galileo when he turned his telescope toward Jupiter, barely more than a point of light, and discovered four little moons in orbit around it! And, at the very same time, how frustrating it must have been to them to know so little about the stars that came out night after night. Remarkably, the ancient Greek Democritus as well as the sixteenth-century astronomer Giordano Bruno proposed that the stars are like our own Sun, but there was no evidence to prove them right. What could they be? What held them in the sky? How far away were they? Why were some brighter than others? Why did they have different colors? And what was that wide band of light reaching from one horizon to the other on a clear night?

The story of astronomy and astrophysics since those days has been the quest to answer those questions, and the additional questions that arose when we started to come up with some answers. For the last four hundred years or so, what astronomers have been able to see, of course, has depended on the power and sensitivity of their telescopes. The great exception was Tycho Brahe, who made very detailed observations with the naked eye that allowed Kepler to arrive at three major discoveries, now known as Kepler's laws.

For most of that time all we had were optical telescopes. I know that sounds odd to a non-astronomer. When you hear "telescope," you think, automatically, "tube with lenses and mirrors that you peer into," right? How could a telescope not be optical? When President Obama hosted an astronomy night in October 2009, there were a bunch of telescopes set up on the White House lawn, and every single one of them was an optical telescope.

But ever since the 1930s, when Karl Jansky discovered radio waves coming from the Milky Way, astronomers have been seeking to broaden the range of electromagnetic radiation through which they observe the universe. They have hunted for (and discovered) microwave radiation (high-frequency radio waves), infrared and ultraviolet radiation (with frequencies just below and just above those of visible light), X-rays, and gamma rays. In order to detect this radiation, we've developed a host of specially designed telescopes—some of them X-ray and gamma ray satellites—enabling us to see more deeply and broadly into the universe. There are even *neutrino* telescopes underground, including one being built right now at the South Pole, called, appropriately enough, IceCube.

For the last forty-five years—my life in astrophysics—I have been working in the field of X-ray astronomy: discovering new X-ray sources and developing explanations for the many different phenomena we observe. As I wrote earlier, the beginning of my career coincided with the heady and exciting early years of the field, and I was in the thick of things for the next four decades. X-ray astronomy changed my life, but more important, it changed the face of astronomy itself. This chapter and the four that follow will take you on a tour of the X-ray universe, from the standpoint of someone who's worked and lived in that universe for his entire scientific career. Let's start with X-rays themselves.

What Are X-rays?

X-rays have an exotic-sounding name, which they received because they were "unknown" (like the x in an equation), but they are simply photons—electromagnetic radiation—making up the portion of the electromagnetic spectrum that we cannot see between ultraviolet light and gamma rays. In Dutch and in German they are not called X-rays; instead they are named after the German physicist, Wilhelm Röntgen, who discovered them in 1895. We distinguish them the same way we identify other inhabitants of that spectrum, in three different but connected ways: by frequency (the number of cycles per second, expressed in hertz), by wavelength (the length of an individual wave, in meters, in this case nanometers), or by energy (in electron volts, eV, or thousands of electron volts, keV).

Here are some quick points of comparison. Green light has a wavelength of about 500 billionths of a meter, or 500 nanometers, and an energy of about 2.5 electron volts. The lowest-energy X-ray photon is about 100 eV, forty times the energy of a photon of green light, with a wavelength of about 12 nanometers. The highest-energy X-rays are about 100 keV, with wavelengths of about 0.012 nanometers. (Your dentist uses X-rays up to about 50 keV.) At the other end of the electromagnetic spectrum, in the United States, radio stations broadcast in the AM band between 520 kilohertz (wavelength 577 meters—about a third of a mile) and 1,710 kilohertz (wavelength 175 meters—nearly twice the length of a football field). Their energy is a billion times less than green light, and a trillion times less than X-rays.

Nature creates X-rays in a number of different ways. Most radioactive atoms emit them naturally during nuclear decay. What happens is that electrons jump down from a higher energy state to a lower one; the difference in energy can be emitted as an X-ray photon. These photons have very discrete energies as the energy levels of the electrons are quantized. Or, when electrons pass by atomic nuclei at high speeds, they change direction and emit some of their energy in the form of X-rays. We call this kind of X-ray emission, which is very common in astronomy as well as in any medical or dental X-ray machine, a difficult German name, bremsstrahlung, which literally means "braking radiation." There are some helpful animated versions of bremsstrahlung X-ray production here: www.youtube.com/watch?v=3fe6rHnhkuY. X-rays of discrete energies can also be produced in some medical X-ray machines, but in general the bremsstrahlung (which produces a continuous X-ray spectrum) dominates. When high-energy electrons spiral around magnetic field lines, the direction of their speed changes all the time and they will therefore also radiate some of their energy in the form of X-rays; we call this synchrotron radiation, but it's also called magnetic bremsstrahlung (this is what is happening in the Crab Nebula—see below).

Nature also creates X-rays when it heats dense matter to very, very high temperatures, millions of degrees kelvin. We call this blackbody radiation (see chapter 14). Matter only gets this hot in pretty extreme circumstances, such as supernova explosions—the spectacular death explosions of some massive stars—or when gas falls at very high speeds toward a black hole or neutron star (more on that in chapter 13, promise!). The Sun, for instance, with a temperature of about 6,000 kelvin at its surface, radiates a little less than half its energy (46 percent) in visible light. Most of the rest is in infrared (49 percent) and ultraviolet (5 percent) radiation. It's nowhere near hot enough to emit X-rays. The Sun does emit some X-rays, the physics of which is not fully understood, but the energy emitted in X-rays is only about one-millionth of the total energy it emits. Your own body emits infrared radiation (see chapter 9); it's not hot enough to emit visible light.

One of the most interesting—and useful—aspects of X-rays is that certain kinds of matter, like bones, absorb X-rays more than others, like soft tissue, which explains why an X-ray image of your mouth or hand shows light and dark areas. If you've had an X-ray, you've also had the experience of being draped with a lead apron to protect the rest of your body, since exposure to X-rays can also increase your risk of getting cancer. Which is why it's mostly a good thing that our atmosphere is such a good absorber of X-rays. At sea level about 99 percent of low-energy X-rays (at 1 keV) are absorbed by just 1 centimeter of air. For X-rays at 5 keV, it takes about 80 centimeters of air, nearly three feet, to absorb 99 percent of the X-rays. For high-energy X-rays at 25 keV, it takes about 80 meters of air to absorb the same proportion.

The Birth of X-ray Astronomy

Now you understand why, back in 1959, when Bruno Rossi had the idea to go looking for X-rays from outer space, he proposed using a rocket that could get completely outside the atmosphere. But his idea about looking for X-rays was wild. There really were no sound theoretical reasons to think there were X-rays coming from outside the solar system. But Rossi was Rossi, and he convinced his former student Martin Annis at American Science and Engineering (AS&E) and one member of his staff, Riccardo Giacconi, that the idea was worth pursuing.

Giacconi and his co-worker Frank Paolini developed special Geiger-Müller tubes that could detect X-rays and fit into the nose cone of a rocket. In fact, they put three of them in one rocket. They called them large-area detectors, but large in those days meant the size of a credit card. The AS&E guys went looking for funding to underwrite this experiment, and NASA turned their proposal down.

Giacconi then changed the proposal by including the Moon as a target and resubmitted it to the Air Force Cambridge Research Laboratories (AFCRL). The argument was that the solar X-rays should produce so-called fluorescent emission from the lunar surface and that this would facilitate chemical analysis of the lunar surface. They also expected bremsstrahlung from the lunar surface due to the impact of electrons present in the solar wind. Since the Moon is so close, X-rays might be detectable. This was a very smart move, as AS&E had already received support from the Air Force for several other projects (some of which were classified), and they may have known that the agency was interested in the Moon. In any event, this time the proposal was approved.

After two rocket failures in 1960 and 1961, the launch one minute before midnight on June 18, 1962, had the stated mission of trying to detect X-rays from the Moon and to search for X-ray sources beyond the solar system. The rocket spent just six minutes above the 80-kilometer mark (over 250,000 feet up), where the Geiger-Müller tubes could detect X-rays in the range from about 1.5–6 keV without atmospheric interference. That's the way you observed in space with rockets in those days. You sent the rockets out of the atmosphere, where they scanned the skies for only five or six minutes, then they came back down.

The truly amazing thing is that right away they found X-rays—not from the Moon, but from someplace outside the solar system.

X-rays from deep space? Why? No one understood the finding. Before that flight we had known of exactly one star that emitted X-rays, our own Sun. And if the Sun had been 10 light-years away, say, which is really just around the corner in astronomical terms, the equipment in that historic flight was a *million* times too insensitive to detect its X-rays. Everyone knew this. So wherever this source was located, it had to emit at least a million times more X-rays than the Sun—and that was only if it was really close by. Astronomical objects that produced (at least) a million or a billion times more X-rays than the Sun were literally unheard of. And there was no physics to describe such an object. In other words, it had to be a brand new kind of phenomenon in the heavens.

A whole new field of science was born the night of June 18–19, 1962: X-ray astronomy.

Astrophysicists began sending up lots of rockets fitted with detectors to figure out precisely where the source was located and whether there were any others. There is always uncertainty in measuring the position of objects in the heavens, so astronomers talk about an "error box," an imaginary box pasted on the dome of the sky whose sides are measured in degrees, or arc minutes, or arc seconds. They make the box big enough so there is a 90 percent chance that the object is really inside it. Astronomers obsess about error boxes, for obvious reasons; the smaller the box, the more accurate the position of the object. This is especially important in X-ray astronomy, where the smaller the box, the more likely it is that you will be able to find the source's optical counterpart. So making the box really, really small is a major achievement.

Professor Andy Lawrence at the University of Edinburgh writes an astronomy blog called The e-Astronomer on which he once posted a reminiscence of working on his thesis, staring at hundreds of position plots of X-ray sources. "One night I dreamt I was an error box, and couldn't find the X-ray source I was supposed to enclose. I woke up sweating." You can understand why!

The size of the error box of the X-ray source discovered by Riccardo Giacconi, Herb Gursky, Frank Paolini, and Bruno Rossi was about 10 degrees × 10 degrees, or 100 square degrees. Now keep in mind that the Sun is half a degree across. The uncertainty in figuring out where the source was consisted of a box the area of which was the equivalent of 500 of our Suns! The error box included parts of constellations Scorpio and Norma, and it touched the border of the constellation Ara. So clearly they were unable to determine in which constellation the source was located.

In April 1963 Herbert Friedman's group at the Naval Research Laboratory in Washington, D.C. improved substantially on the source's position. They found that it was located in the constellation Scorpio. That's why the source is now known as Sco X-1. The X stands for "X-rays," and the 1 indicates that it was the first X-ray source discovered in the constellation Scorpio. It is of historical interest, though never mentioned, that the position of Sco X-1 is about 25 degrees away from the center of the error box given in the Giacconi et al. paper that marked the birth of X-ray astronomy. When astronomers discovered new sources in the constellation Cygnus (the Swan), they received the names Cygnus X-1 (or Cyg X-1 for short), Cyg X-2, and so on; the first source discovered in the constellation Hercules was Her X-1; in Centaurus Cen X-1. Over the next three years about a dozen new sources were discovered using rockets, but with one important exception, namely Tau X-1, located in the constellation Taurus, no one had any idea what they were, or how they were producing X-rays in such huge quantities that we could detect them thousands of light-years away.

The exception was one of the more unusual objects in the sky: the Crab Nebula. If you don't know about the Crab Nebula, it's worth turning to the photo insert to look at the image of it there now—I suspect you'll recognize it right away. There are also many photos of it on the web. It's a truly remarkable object about 6,000 light-years away—the stunning remains of a supernova explosion in the year 1054 recorded by Chinese astronomers (and quite possibly in native American pictographs—take a look here: http://seds.org/messier/more/m001_sn.html#collins1999) as a superbright star in the heavens that suddenly appeared, more or less out of nowhere, in the constellation Taurus. (There is some disagreement about the exact date, though many claim July 4.) That month it was the brightest object in the sky other than the Moon; it was even visible during the day for several weeks, and you could still see it at night for another two years.

Once it faded, however, scientists apparently forgot about it until the eighteenth century, when two astronomers, John Bevis and Charles Messier, found it independently of each other. By this time, the remains of the supernova (called a supernova remnant) had become a nebular (cloudlike) object. Messier developed an important astronomical catalog of objects like comets, nebulae, and star clusters—the Crab Nebula is the first object in his catalog, M-1. In 1939 Nicholas Mayall from Lick

Observatory (in Northern California) figured out that M-1 is the remnant of the supernova of 1054. Today, a thousand years after the explosion, there is still such wonderful stuff going on inside the Crab Nebula that some astronomers devote entire careers to studying it.

Herb Friedman's group realized that the Moon was going to pass right in front of the Crab Nebula on July 7, 1964, and block it from view. The term astronomers use for this blocking out is "occultation"—that is, the Moon was going to occult the Crab Nebula. Not only did Friedman want to confirm that the Crab Nebula was indeed an X-ray source, but he also was hoping he could demonstrate something else—something even more important.

By 1964 a renewed interest had emerged among astronomers in a type of stellar object whose existence was first postulated during the 1930s but that had never been detected: neutron stars. These strange objects, which I discuss more fully in chapter 12, had been conjectured to be one of the final stages in a star's life, possibly born during a supernova explosion and composed mostly of neutrons. If they existed, they would be of such great density that a neutron star with the mass of our Sun would only be about 10 kilometers in diameter—about 12 miles all the way across, if you can imagine such a thing. In 1934 (two years after the discovery of neutrons), Walter Baade and Fritz Zwicky had coined the term "supernova" and proposed that neutron stars might be formed in supernova explosions. Friedman thought that the X-ray source in the Crab Nebula might be just such a neutron star. If he was right, the X-ray emission he was seeing would disappear abruptly when the Moon passed in front of it.

He decided to fly a series of rockets, one after the other, right as the Moon was going in front of the Crab Nebula. Since they knew the Moon's exact position in the sky as it moved, and could point the counters in that direction, they could "watch" for a decline in X-rays as the Crab Nebula disappeared. As it happened, their detectors did indeed pick up a decline, and this observation was the first conclusive optical identification of an X-ray source. This was a major result, since once we had made an optical identification, we were optimistic that we would soon discover the mechanism behind these enigmatic and powerful X-ray sources.

Friedman, however, was disappointed. Instead of "winking out" as the Moon passed over the Crab Nebula, the X-rays disappeared gradually, indicating that they came from the nebula as a whole and not from a single small object. So he hadn't found a neutron star. However, there *is* a very special neutron star in the Crab Nebula, and it *does* emit X-rays; the neutron star rotates about its axis about thirty times per second! If you want a real treat, go to the Chandra X-Ray Observatory website (http://chandra.harvard.edu/) and call up images of the Crab Nebula. I promise you, they are stunning. But forty-five years ago we had no orbiting imaging X-ray telescopes in space, so we had to be much more inventive. (After the 1967 discovery of radio pulsars by Jocelyn Bell, in 1968 Friedman's group finally detected X-ray pulsations—about thirty per second—from the neutron star in the Crab Nebula.)

Just as Friedman was observing the occultation of the Crab, my friend (to be) George Clark at MIT was in Texas preparing for a high-altitude balloon night flight to search for high-energy X-rays from Sco X-1. But when he heard about Friedman's results—even without the Internet, news traveled fast—he completely changed his plans and switched to a day flight in search of X-rays in excess of about 15 keV from the Crab Nebula. And he found them too!

It's hard to put into words just how exciting all this was. We were at the dawn of a new era in scientific exploration. We felt we were lifting a curtain that had been hiding these amazing realms of the universe. In reality, by getting our detectors up so high, by getting into space, by getting to the top of the atmosphere where X-rays could penetrate without being absorbed by air, we were removing blinding filters that had been on our eyes for all of human history. We were operating in a whole new spectral domain.

That has happened often in the history of astronomy. Every time we learned that objects in the heavens emitted new or different kinds of radiation, we had to change what we thought we knew about stars, about their life cycles (how they are born, how they live, and how and why they die), about the formation and evolution of clusters of stars, about galaxies, and even about clusters of galaxies. Radio astronomy, for instance, showed us that the centers of galaxies can emit jets hundreds of thousands of light-years long; it has also discovered pulsars, quasars, and radio galaxies and is responsible for the discovery of cosmic microwave background radiation, which radically changed our views of the early universe. Gamma-ray astronomy has discovered some of the most powerful and (fortunately) distant explosions in the universe, known as gamma-ray bursts, which emit afterglows in X-rays and visible light all the way down to radio waves.

We knew that the discovery of X-rays in space was going to change our understanding of the universe. We just didn't know how. Everywhere we looked with our new equipment, we saw new things. That's not surprising, perhaps. When optical astronomers started getting images from the Hubble Space Telescope, they were thrilled, awestruck, and—maybe this isn't so obvious—hungry for more. But they were basically extending the reach of a centuries-old instrument, in a field dating back millennia. As X-ray astronomers, we were experiencing the dawn of a whole new scientific field. Who knew where it would lead, or what we would discover? We surely didn't!

How fortunate for me that Bruno Rossi invited me to MIT in January 1966, just as this field was taking off, and that I immediately joined George Clark's group. George was a very, very smart physicist, a really impressive guy with whom I became friends for the rest of my life. Even now, I can hardly believe my good luck—a great friend and a new career, both in the same month.

CHAPTER 11

X-ray Ballooning, the Early Days

W hen I arrived at MIT, there were five active balloon groups in the world: George Clark at MIT, Ken McCracken at the University of Adelaide in Australia, Jim Overbeck at MIT, Larry Peterson at UC San Diego, and Bob Haymes at Rice University. This chapter is largely about my own experiences with X-ray ballooning, which was at the center of my research in the decade between 1966 and 1976. During these years I made observations from Palestine, Texas; Page, Arizona; Calgary, Canada; and Australia.

Our balloons carried our X-ray detectors to an altitude of about 145,000 feet (about 30 miles), where the atmospheric pressure is only 0.3 percent of that at sea level. When the atmosphere is this thin, a good fraction of X-rays with energies above 15 keV get through.

Our balloon observations complemented the rocket observations. Rocket-borne detectors typically observed X-rays in the range from 1 to 10 keV and only for about five minutes during an entire flight. Balloon observations could last for hours (my longest flight was twenty-six hours) and my detectors observed X-rays in the range above 15 keV.

Not all sources that were detected during rocket observations were

detectable during balloon observations, since the sources often emitted most of their energy at low-energy X-rays. On the other hand, we
were able to detect sources emitting largely high-energy X-rays invisible
during rocket observations. Thus, not only did we discover new sources
and extend the spectra of known sources to high energies, but we also
were capable of detecting variability in the X-ray luminosity of sources
on time scales of minutes to hours, which was not possible with rockets.
This was one of the early successes of my research in astrophysics.

In 1967 we discovered an X-ray flare from Sco X-1—that was a real
shocker—I'll tell you all about this later in this chapter. My group also
discovered three X-ray sources, GX 301-2, GX 304-1, and GX 1+4, never
seen before during rocket observations, and all of them showed changes
in their X-ray intensity on time scales of minutes. GX 1+4 even showed
periodic variability with a period of about 2.3 minutes. At the time we
had no idea what could be the cause of such rapid changes in the X-ray
intensity, let alone the 2.3-minute periodicity, but we knew we were
breaking new ground—uncovering new territory.

For some astronomers, though, even in the late 1960s, the significance of X-ray astronomy hadn't yet sunk in. In 1968, I met the Dutch
astronomer Jan Oort at Bruno Rossi's home. Oort was one of the most
famous astronomers. He had been an incredible visionary; right after
World War II, he started a whole radio astronomy program in the Netherlands. When he came to MIT that year, I showed him the balloon data
from our flights in 1966 and 1967. But he said to me—and I'll always
remember this—"X-ray astronomy is just not very important." Can
you believe it? "Just not very important." He couldn't have been more
wrong. This was one of the greatest astronomers of all time, and he was
completely blind to its significance. Maybe because I was younger, and
hungrier—to be fair, he was sixty-eight by then—it was obvious to me
that we were harvesting pure gold, and we were only just scratching the
surface.

I remember in the 1960s and 1970s I would read every single paper
that came out on X-ray astronomy. In 1974 I gave five lectures in Leiden

(Oort was in my audience), and I was able to cover *all* of X-ray astronomy. Nowadays thousands of papers on X-ray astronomy are published every year, in a multitude of subfields, and no one can grasp the entire field. Many researchers spend their entire careers on one of dozens of specific topics such as single stars, accretion disks, X-ray binaries, globular clusters, white dwarfs, neutron stars, black holes, supernovae remnants, X-ray bursts, X-ray jets, galactic nuclei, and clusters of galaxies. The early years were the most fantastic years for me. They were also demanding, in just about every way: intellectually, physically, even logistically. Launching balloons was so complicated and expensive, time-consuming, and tension producing, I can hardly describe it. I'll try, though.

Getting Aloft: Balloons, X-ray Detectors, and the Launch

Before a physicist can do anything (unless, that is, you're a theorist, who may need only a piece of paper or a computer screen), you have to get the money to build equipment and pay students and sometimes to travel very far. Lots of what scientists really do is write grant proposals, in highly competitive programs, to get supported to do research. I know it's not sexy or romantic, but believe me, nothing happens without it. Nothing.

You could have a wonderful idea for an experiment or an observation, and if you don't know how to transform it into a winning proposal, it goes nowhere. We were always competing against the best in the world, so it was a cutthroat business. It still is, for just about any scientist in any field. Whenever you look at a successful experimental scientist—in biology, chemistry, physics, computer science, economics, or astronomy, it doesn't matter—you are also looking at someone who's figured out how to beat the competition over and over again. That does not make for warm and fuzzy personalities, for the most part. It's why my wife, Susan, who's worked at MIT for ten years, is fond of saying, "There are no small egos at MIT."

Suppose we got the funding, which we usually did (I was gener-ously supported by the National Science Foundation and NASA). To send a balloon up nearly 30 miles, carrying a 2,000-pound X-ray tele-scope (connected to a parachute), which you had to recover intact, was a very complex process. You had to have reliably calm weather at launch, because the balloons were so delicate that a gust of wind could sink the whole mission. You needed to have some infrastructure—launch sites, launch vehicles, and the like—to help get the balloons way up into the atmosphere and to track them. Since I wanted to observe in the general direction of the center of the Milky Way, which we call the galactic cen-ter, where many X-ray sources were located, I needed to observe from the Southern Hemisphere. I chose to launch from Mildura and Alice Springs, Australia. I was very far away from my home and family—I had four children by then—usually for a couple of months at a time.

Everything about the launches was expensive. The balloons them-selves were enormous. The largest one I flew (which at the time was the largest balloon ever flown, and it may well still be the largest ever) had a volume of 52 million cubic feet; when fully inflated and flying at 145,000 feet, its diameter was about 235 feet. The balloons were made of very lightweight polyethylene—one-half of one-thousandth of an inch thick, thinner than Saran Wrap or cigarette paper. If they ever touched the ground during launch, they would tear. These gigantic, beautiful bal-loons weighed about 700 pounds. We usually traveled with a backup, and each one cost $100,000—forty years ago, when that was real money.

They had to be made in immense plants. The gores, the sections of the balloon that look like tangerine skin segments, were made separately and then put together by heat sealing. The manufacturer only trusted women to do the sealing; they said it was well known that men were too impa-tient and made too many mistakes. Then we had to ship the helium to inflate the balloons all the way to Australia. The helium alone cost about $80,000 per balloon. In current dollars that was more than $700,000 for just one balloon and its helium, without even considering the backup balloon, our transportation, lodging, or food. That's right—here we were trying to ferret out the secrets of deep space, living in the middle of the Australian desert, utterly dependent on the weather. And I haven't even told you about Jack. I'll get to Jack in a bit.

But the balloons were cheap compared to the telescopes. Each telescope, an extremely complicated machine weighing about a ton, took roughly two years to build and cost $1 million—$4 million in today's dollars. We never had enough money for two telescopes at a time. So if we lost our telescope—which happened twice—we were out of luck for at least two years. We couldn't even start building a new one until we'd gotten the funding. So it was a catastrophe if we lost one.

And not just for me, not at all. This would cause a major delay for my graduate students, who were all deeply involved in building the telescopes, and whose PhD theses were about the instruments and the results of the observations. Their degrees went up in the air with the balloons.

We needed the cooperation of the weather, too. There are intense winds in the stratosphere, flowing east to west at about 100 miles per hour for about six months of the year, and west to east the other half of the year. Twice a year these winds reverse direction—we call it the turnaround—and as they do, the wind speeds at 145,000 feet become very low, which would allow us to make observations for many hours. So we needed to be in a place where we could measure these winds and could launch during the turnaround. We probed every other day with weather balloons that we tracked by radar. Most of the time they would make it to about 125,000 feet, about 24 miles up, before they popped. But predicting the atmosphere isn't like pushing ball bearings down a track in a lab demonstration. The atmosphere is so much more complex, so much less predictable, and yet everything we did depended on making good forecasts.

There was more. At an altitude between about 30,000 and 60,000 feet the atmosphere is called the tropopause, where it's very, very cold—minus 50 degrees Celsius (–58°F)—and our balloons would get very brittle. There were jet stream winds too, and they beat on the balloon, which could then burst. So many things could go wrong. Once my balloon blew out to sea—end of telescope. The payload was found nine months later on a beach in New Zealand. Miraculously, with the help of Kodak, we were able to retrieve the data, which were recorded on film on board.

We prepared over and over and over for these launches, and yet I always said that even though we prepared like crazy, we still needed a little luck. Sometimes a lot of luck. We would bring the equipment to this remote station. Then we did tests on the telescope, calibrating the instruments and making sure everything was working. We would go through the rigging connecting the telescope to the parachute, which would eventually connect to the balloon as well. It could take us about three weeks to do all the tests at the balloon launching site and be flight ready, and then the weather might not cooperate. And we had nothing else to do then except to sit there and wait and keep the batteries charged. It's a good thing Alice Springs was so beautiful: a fantastic desert town right in the heart of Australia. It really felt like it was in the middle of nowhere, but the skies were clear and the early mornings when we tried to launch were spectacular: the night sky had turned its predawn deep blue, and as the Sun rose it painted the sky and the desert in brilliant pinks and oranges.

Once we were ready to go, we needed to have winds under 3 miles per hour in a steady direction for three or four hours, which is how long it took to get the balloon off the ground (the inflation alone took two hours). That's why we mostly launched at dawn, when there was the least amount of wind. But it could happen that our forecast was wrong, and we just had to wait, and wait, and wait some more, until the weather cooperated.

We were in the middle of a launch one time in Mildura—we had not even started inflation—and the wind came up, contrary to the weatherman's forecast. The balloon was destroyed, but thank goodness the telescope was safe! All that preparation, and $200,000—gone in a few seconds. Talk about painful. All we could do was wait for better weather and try again with our spare balloon.

The failures stick with you. On my last expedition to Alice Springs we lost two balloons in a row right at launch, because the launch crew made some tragic mistakes. Our expedition was a complete failure—but at least our telescope wasn't damaged. It never left the ground. On my last expedition (in 1980), in Palestine, Texas, the eight-hour flight was successful, but when we terminated the flight by radio command, we lost our telescope; the parachute never opened.

Even today, balloon launches are far from a sure thing. In an attempted NASA launch from Alice Springs in April 2010, something went wrong and the balloon crashed while trying to take off, destroying millions of dollars worth of equipment and nearly injuring onlookers. You can see the story here: www.physorg.com/news191742850.html.

Over the years I must have launched about twenty balloons. I had only five that failed during launch or didn't get to altitude (they may have been leaking helium). That was considered a good success rate (75 percent). In the insert you can see a picture of the inflation (with helium) of a balloon and also a picture of a balloon launch.

Months before going to the launch site, we would test the payload at a firm in Wilmington, Massachusetts. We put the telescope into a vacuum chamber and brought the air pressure down to the same we'd have way up high, about three-thousandths of an atmosphere. Then we cooled it down to minus 50 degrees Celsius (–58°F) and ran it—turning on all the X-ray detectors and monitoring for ten seconds every twenty minutes X-rays from a radioactive source for twenty-four hours straight. Some of our competitors' telescopes—yes, we did feel like the other teams doing the same kinds of things were our competition—would fail sometimes because their batteries would lose power at low temperatures or quit altogether. That never happened to us because we had tested them so thoroughly. If we saw in the testing period that our batteries were going to lose power, we figured out how to heat them up if necessary and keep the power going.

Or take the problem of corona discharge—sparking from high-voltage wires. Some of our equipment ran on high voltage, and very thin air, where the pressure is very low, is an ideal environment for sparks, from wires into the open air. Remember the buzz around transmission lines I mentioned back in chapter 7? That's corona discharge. Every experimental physicist who works with high voltage knows you can get corona discharge. I show examples of these sparks in my classes. There, corona discharge is fun. At 145,000 feet, it's a catastrophe.

In lay terms, the equipment would start to sputter, and you would get so much electronic noise that you couldn't pick out the X-ray photons. How big a disaster would this be? Total and complete: you would get no usable data at all on a flight. The solution was to coat all of our high-voltage wires in silicon rubber. Other folks did the same thing and still got corona discharge. Our testing and preparation paid off. We *never* had corona discharge. This was just one of dozens of complex engineering issues involved in building these intricate telescopes—that's why they took so long to build, and cost so much money.

So, once we got the telescope high up into the atmosphere, how did we detect X-rays? The answer to this question is not simple, so please bear with me. To begin with, we used a special kind of detector (sodium iodide crystals), not the proportional counters (filled with gas) the rockets used, but something that was able to detect X-rays with energies higher than 15 keV. When an X-ray photon penetrates one of these crystals it can kick an electron out of its orbit and transfer its X-ray energy to that electron (this is called photoelectric absorption). This electron in turn will produce a track of ions in the crystal before it comes to a stop. When these ions get neutralized, they release energy mostly in the form of visible light; thus a flash of light is produced—the energy of the X-ray photon is converted into a light flash. The higher the energy of the X-rays, the stronger the light flashes. We used photomultipliers to detect the light flashes and convert them into electric pulses: the brighter the light flash, the higher the voltage of a pulse.

We then amplified these pulses and sent them to a discriminator, which measured the voltage of the electric pulses and sorted them according to magnitude—which indicated the energy levels of the

X-rays. In the early days we recorded the X-rays at only five different energy levels.

So that we would have a record of the detections after the balloon flight, in the early days we recorded them on board, by energy level and the time they were detected. We wired the discriminator to send these sorted impulses to light-emitting diodes, which created a pattern of flashing lights at those five distinct energy levels. Then we photographed those flashing lights with a camera running continuous film.

If a light was on, it would make a track on the film. All together, the film of an observation would look like a series of dashes and lines, lines and dashes. Back at MIT we would "read" the film with a special reader designed by George Clark that converted the lines and dashes to punch tape: paper tape with holes in it. Then we read the punch tape with light sensitive diodes and recorded the data on magnetic tape. We had written a program on computer cards in Fortran (I realize this sounds prehistoric) and used it to read the magnetic tape into the memory of the computer, which—finally!—gave us X-ray counts as a function of time in the five different energy channels.

I know it sounds like a Rube Goldberg machine. But think about what we were trying to do. We were trying to measure the counting rate (the number of X-rays per second) and energy levels of X-ray photons, as well as the location of the source that had emitted them—photons that had been traveling for thousands of years at the speed of light, spreading through the galaxy and thinning out continuously by the square of the distance they traveled. And unlike a stable mountaintop optical telescope whose control system can keep the telescope trained on the same spot for many hours and can return to the same spot night after night, we had to make use of whatever time we had (at most once per year)—always measured in hours—while a fragile balloon carried our thousand-kilo telescope 145,000 feet above the Earth.

When a balloon was in flight I followed it in a small plane, usually keeping it in sight (in the daytime, that is—not at night), flying at just 5,000 or 10,000 feet. You can imagine what that was like, for many hours at a time. I'm not a small man. It was easy, all too easy, to get sick in these little four-seater planes, flying for eight, ten, twelve hours at a time. Plus, I was nervous the whole time the balloon was up. The only time you could relax was after the recovery, when you had the data in hand.

The balloon was so enormous that even though it was nearly 30 miles up, when sunlight hit it, you could see it very clearly. With radar, we could follow it a long way from the launching station until the curvature of the Earth would make that impossible. That's why we outfitted the balloon with a radio transmitter, and at night we had to switch exclusively to tracking the balloon by radio beacon. No matter how hard we worked getting articles in the local newspapers about the launch, the balloons could drift hundreds of miles, and when they were aloft we'd get all kinds of reports of UFOs. It was funny, but it made perfect sense, really. What else were people supposed to think when they caught a glimpse of a mysterious entity in the sky of indeterminate size and distance? To them it really was an unidentified flying object. You can see a picture taken with a telescope of a balloon at 145,000 feet in the insert.

Even with all our planning, and weather forecasts, and even in turnaround, the winds at 145,000 feet altitude could turn out to be unreliable. Once, in Australia, we had expected the balloon to head north from Alice Springs, but instead it took off straight south. We followed it visually until sunset and kept radio contact with it through the night. By morning it was getting too close to Melbourne, and we were not allowed to enter the air space between Sydney and Melbourne. No one was going to shoot it down, but we had to do something. So when our wayward balloon was just about to reach forbidden air space, we reluctantly gave the radio command that cut the payload loose. Separating the telescope from the balloon would shatter the balloon—it could not survive the shock wave caused by the sudden release of the payload—and the telescope would start to fall, the parachute would open (except in 1980) and slowly float down, bringing the telescope safely back to Earth. Huge pieces of the balloon would also hit the ground, usually spread out over an acre or more. This occurred sooner or later in every balloon flight, and it was always a sad moment (even though it was always necessary), because we were terminating the mission—cutting off the data flow. We wanted the telescope to be aloft as long as possible. We were so hungry for data in those days—that was the whole point.

Recovery in the Outback: Kangaroo Jack

We put cardboard crash pads on the bottom of the telescope to soften its landing. If it was during the day, and we had visual contact with the balloon (which would suddenly disappear when we sent the cut-down command), we would soon spot the parachute; we did our best to follow it all the way down, circling it in our little airplane. Once it landed we would mark its location on a very detailed map as accurately as possible.

Then the really bizarre part started: because here we were, in an airplane, and our payload, with all our data, the culmination of years of work, was lying on the ground, almost within reach, but we couldn't just land in the middle of the desert and get it! What we had to do was to draw the attention of local people, and the way we usually did this was by flying a plane low over a house. Houses were pretty far apart in the desert. Residents knew what the low-flying plane meant and usually came out of the house and made contact by waving. Then we would land at the nearest airstrip (not to be confused with an airport) in the desert and wait for them to show up.

During one flight, there were so few houses in the area that we had to hunt for a while. Eventually we found this guy Jack living in the desert 50 miles away from his nearest neighbor. He was drunk and pretty crazy. We didn't know that at first, of course. But we made contact from the air and then flew to the airstrip and waited; after about 15 hours he showed up with his truck, a battered old jeep-like thing with no windshield, just a roof on its cab, and an open bay in back. Jack liked to tear around the desert at 60 miles an hour, chasing and shooting kangaroos.

I stayed with Jack and the truck and one of my graduate students, while our tracking airplane directed us to the payload. The truck needed to go across unmarked terrain. We kept in radio contact with the plane. We were lucky with Jack. From all that kangaroo hunting he really knew where he could drive.

He also had this awful game I hated, but we were already depending on him, so there wasn't much I could do; he gave me a demonstration just once. He put his dog on the roof of the jeep, accelerated up to 60 miles an hour, then slammed on the brakes, and the dog catapulted through the air onto the ground. The poor dog! Jack laughed and laughed and then delivered his punch line: "You can't teach an old dog new tricks."

It took us half a day to reach the payload, which was being guarded by a six-foot-long iguana—a really nasty-looking creature. To tell the truth, it scared the hell out of me. But of course I didn't want to show that, so I said to my graduate student, "There's no problem. These animals are harmless. You go first." And he did, and it turns out that they *are* harmless, and during the entire four hours it took us to recover the payload and get it on Jack's truck, this animal never moved.

The Balloon Professor

Then we went back to Alice Springs, and of course we were on the front page of the *Centralian Advocate* with a great photograph of the balloon launch. The headline read START OF SPACE PROBE and the article talked about the "balloon professor" having returned. I had become a sort of local celebrity and gave talks to the Rotary Club and for students at the high school, even once in a steak house, which earned me dinner for my crew. What we really wanted to do was take our film back home as quickly as possible, develop and analyze it, and see what we'd found. So after a few days' cleanup we were on our way. You can see just how demanding this kind of research was. I was away from home for something like two months at least every other year (sometimes every year). And there's no question about it that my first marriage suffered a lot because of it.

At the same time, despite all the nervousness and tension, it was exciting and great fun and I was proud of my graduate students, notably Jeff McClintock and George Ricker. Jeff is now senior astrophysicist at the Harvard-Smithsonian Center for Astrophysics and won the 2009 Rossi Prize (named for guess who?) for his work measuring the masses of black holes in X-ray binary star systems. (We'll get to that in chapter 13.) George, I'm happy to say, still works at MIT. He is brilliant at designing and developing innovative new instrumentation. He is best known for his research in gamma-ray bursts.

Ballooning was very romantic in its way. To be up at four o'clock in the morning, drive out to the airport, and see the sunrise and see the spectacular inflation of the balloon—this beautiful desert, under the sky, just stars at first, and then slowly seeing the Sun come up. Then, as the balloon was released and pulled itself into the sky, it shimmered silver and gold in the dawn. And you knew just how many little things had to go just right, so all your nerves were jangling the entire time. My goodness. And if it seemed to be a good launch, in which the myriad details (each one of which was the source of a potential disaster) seemed to fall into place one after another—what an incredible feeling!

We really were on the cutting edge in those days. To think that success partly depended on the generosity of a drunken Australian kangaroo hunter.

An X-ray Flare from Sco X-1

No discovery we made in those years was more thrilling for me than the totally unexpected finding that some X-ray sources have remarkable flare-ups in the amount of X-rays they emit. The idea that the X-ray intensity from some sources varies was in the air as early as the mid-1960s. Philip Fisher and his group at Lockheed Missiles and Space Company compared the X-ray intensities of seven X-ray sources detected during their rocket flight on October 1, 1964, with those of a rocket flight by Friedman's group on June 16, 1964. They found that the X-ray intensity (which we call X-ray flux) for the source Cyg XR-1 (now called Cyg

X-1) was five times lower on October 1 than on June 14. But whether or not this observation demonstrated real variability was unclear. Fisher's group pointed out that the detectors used by Friedman's group were much more sensitive to low-energy X-rays than the detectors they had used and that this might explain the difference.

The issue was settled in 1967 when Friedman's group compared the X-ray flux of thirty sources over the prior two years and determined that many sources really did vary in intensity. Particularly striking was the variability of Cyg X-1.

In April 1967, Ken McCracken's group in Australia launched a rocket and discovered a source nearly as bright as Sco X-1 (the brightest X-ray source we knew of), which had not shown up when detectors had observed the same spot a year and a half earlier. Two days after the announcement of this "X-ray nova" (as it was called) during the spring meeting of the American Physical Society in Washington D.C., I was on the phone with one of the most eminent pioneers in X-ray astronomy, and he said to me, "Do you believe that nonsense?"

Its intensity went down in a few weeks by a factor of three, and five months later its intensity had diminished by at least a factor of fifty. Nowadays, we call these sources by the pedestrian name "X-ray transients."

McCracken's group had located the source in the constellation Crux, which you may know better as the Southern Cross. They were very excited about this, and it became something of an emotional thing for them, since that very constellation is in the Australian flag. When it turned out that the source's location was just outside the Southern Cross, in Centaurus instead, the original name Crux X-1 was changed to Cen X-2, and the Aussies were very disappointed. Scientists can get very emotional about our discoveries.

On October 15, 1967, George Clark and I observed Sco X-1 in a ten-hour balloon flight launched from Mildura, Australia, and we made a major discovery. This discovery wasn't anything like you see in pictures of the NASA Space Center in Houston, where they all cheer and hug one another when they have a success. They are seeing things happen in real time. During our observing we had no access to the data; we were just hoping that the balloon would last and that our equipment would work flawlessly. And, of course, we always worried about how to get the telescope and the data back. That's where all the nerves and the excitement were.

We analyzed our data months later, back home at MIT. I was in the computer room one night, assisted by Terry Thorsos. We had very large computers at MIT in those days. The rooms had to be air-conditioned because the computers generated so much heat. I remember that it was around eleven p.m. If you wanted to get some computer runs, the evening was a good time to sneak in some jobs. In those days you always needed to have a computer operator to run your programs. I got into a queue and waited patiently.

So here I was, looking at the balloon data, and all of a sudden I saw a large increase in the X-ray flux from Sco X-1. Right there, on the printout, the X-ray flux went up by a factor of four in about ten minutes, lasted for nearly thirty minutes, and then subsided. We had observed an X-ray flare from Sco X-1, and it was enormous. *This had never been observed before.* Normally, you'd say to yourself, "Is this flare something that could be explained in a different way? Was it perhaps caused by a malfunctioning detector?" In this case, there was no doubt in my mind. I knew the instrument inside and out. I trusted all our preparation and testing, and throughout the flight we had checked the detector continuously and had measured the X-ray spectrum of a known radioactive source every twenty minutes as a control—the instruments were working flawlessly. I trusted the data 100 percent. Looking at the printout I could see that the X-ray flux went up and down; of all the sources we observed in that ten-hour flight, only one shot up and down, and that was Sco X-1. It was real!

The next morning I showed George Clark the results, and he nearly fell off his chair. We both knew the field well; we were overjoyed! No one had anticipated, let alone observed, a change in the flux of an X-ray source on a time scale of ten minutes. The flux from Cen X-2 decreased by a factor of three within a few weeks after the first detection, but here we had variability by a factor of four within ten minutes—about three thousand times faster.

We knew that Sco X-1 emitted 99.9 percent of its energy in the form of X-rays, and that its X-ray luminosity was about 10,000 times the total luminosity of our Sun and about 10 billion times the X-ray luminosity of the Sun. For Sco X-1 to change its luminosity by a factor of four on a time scale of ten minutes—well, there was simply no physics to understand it. How would you explain it if our Sun would become four times brighter in ten minutes? It would scare the hell out of me.

The discovery of variability on this time scale may have been the most important discovery in X-ray astronomy made from balloons. As I mentioned in this chapter, we also discovered X-ray sources that the rockets couldn't see, and those were important discoveries as well. But nothing else had the impact of Sco X-1's ten-minute variability.

It was so unexpected at the time that many scientists couldn't believe it. Even scientists have powerful expectations that can be difficult to challenge. The legendary editor of the *Astrophysical Journal Letters,* S. Chandrasekhar, sent our Sco X-1 article to a referee, and the referee didn't believe our finding at all. I still remember this, more than forty years later. He wrote, "This must be nonsense, as we know that these powerful X-ray sources cannot vary on a time scale of ten minutes."

We had to talk our way into the journal. Rossi had had to do exactly the same thing back in 1962. The editor of *Physical Review Letters,* Samuel Goudsmit, accepted the article founding X-ray astronomy because Rossi was Rossi and was willing, as he wrote later, to assume "personal responsibility" for the contents of the paper.

Nowadays, because we have instruments and telescopes that are so much more sensitive, we know that many X-ray sources vary on *any* timescale, meaning that if you observe a source continuously day by day, its flux will be different every day. If you observe it second by second it will change as well. Even if you analyze your data millisecond by millisecond you may find variability in some sources. But at the time, the ten-minute variability was new and unexpected.

I gave a talk about this discovery at MIT in February 1968, and I was thrilled to see Riccardo Giacconi and Herb Gursky in the audience. I felt as though I'd arrived, that I had been accepted into the cutting edge of my field.

In the next few chapters I'll introduce you to the host of mysteries that X-ray astronomy solved, as well as to some we astrophysicists are still struggling to find answers for. We'll travel to neutron stars and plunge into the depths of black holes. Hold on to your hats.

CHAPTER 12

Cosmic Catastrophes,
Neutron Stars, and Black Holes

Neutron stars are smack dab at the center of the history of X-ray astronomy. And they are really, really cool. Not in terms of temperature, not at all: they can frequently have surface temperatures upward of a million kelvin. More than a hundred times hotter than the surface of our Sun.

James Chadwick discovered the neutron in 1932 (for which he received the Nobel Prize in Physics in 1935). After this extraordinary discovery, which many physicists thought had completed the picture of atomic structure, Walter Baade and Fritz Zwicky hypothesized that neutron stars were formed in supernova explosions. It turns out that they were right on the money. Neutron stars come into being through truly cataclysmic events at the end of a massive star's lifetime, one of the quickest, most spectacular, and most violent occurrences in the known universe—a core-collapse supernova.

A neutron star doesn't begin with a star like our Sun, but rather with a star at least eight times more massive. There are probably more than a billion such stars in our galaxy, but there are so many stars of all kinds in our galaxy that even with so many, these giants must still be considered rare.

Like so many objects in our world—and universe—stars can only "live" by virtue of their ability to strike a rough balance between immensely powerful forces. Nuclear-burning stars generate pressure from their cores where thermonuclear reactions at temperatures of tens of millions of degrees kelvin generate huge amounts of energy. The temperature at the core of our own Sun is about 15 million kelvin, and it produces energy at a rate equivalent to more than a billion hydrogen bombs per second.

In a stable star, this pressure is pretty well balanced by the gravity generated by the huge mass of the star. If these two forces—the outward thrust of the thermonuclear furnace and the inward-pulling grip of gravity—didn't balance each other, then a star wouldn't be stable. We know our Sun, for example, has already had about 5 billion years of life and should continue on that path for another 5 billion years. When stars are about to die, they really change, and in spectacular ways. When stars have used up most of the nuclear fuel in their cores, many approach the final stages of their lives by first putting on a fiery show. This is especially true for massive stars. In a way, supernovae resemble the tragic heroes of theater, who usually end their overlarge lives in a paroxysm of cathartic emotion, sometimes fiery, often loud, evoking, as Aristotle said, pity and terror in the audience.

The most extravagant stellar demise of all is that of a core-collapse supernova, one of the most energetic phenomena in the universe. I'll try to do it justice. As the nuclear furnace at the core of one of these massive stars begins to wind down—no fuel can last forever!—and the pressure it generates begins to weaken, the relentless, everlasting gravitational attraction of the remaining mass overwhelms it.

This process of exhausting fuel is actually rather complicated, but it's also fascinating. Like most stars, the really massive ones begin by burning hydrogen and creating helium. Stars are powered by nuclear energy—not fission, but fusion: four hydrogen nuclei (protons) are fused together into a helium nucleus at extremely high temperatures, and this produces heat. When these stars run out of hydrogen, their cores shrink (because of the gravitational pull), which raises the temperature high enough that they can start fusing helium to carbon. For stars with masses more than about ten times the mass of the Sun, after carbon burning they go through oxygen burning, neon burning, silicon burning, and ultimately form an iron core.

After each burning cycle the core shrinks, its temperature increases, and the next cycle starts. Each cycle produces less energy than the previous cycle and each cycle is shorter than the previous one. As an example (depending on the exact mass of the star), the hydrogen-burning cycle may last 10 million years at a temperature of about 35 million kelvin, but the last cycle, the silicon cycle, may only last a few days at a temperature of about 3 billion kelvin! During each cycle the stars burn most of the products of the previous cycle. Talk about recycling!

The end of the line comes when silicon fusion produces iron, which has the most stable nucleus of all the elements in the periodic table. Fusion of iron to still heavier nuclei doesn't produce energy; it requires energy, so the energy-producing furnace stops there. The iron core quickly grows as the star produces more and more iron.

When this iron core reaches a mass of about 1.4 solar masses, it has reached a magic limit of sorts, known as the Chandrasekhar limit (named after the great Chandra himself). At this point the pressure in the core can no longer hold out against the powerful pressure due to gravity, and the core collapses onto itself, causing an outward supernova explosion.

Imagine a vast army besieging a once proud castle, and the outer walls begin to crumble. (Some of the battle scenes in the Lord of the Rings movies come to mind, when the apparently limitless armies of Orcs break through the walls.) The core collapses in milliseconds, and the matter falling in—it actually races in at fantastic speeds, nearly a quarter the speed of light—raises the temperature inside to an unimaginable 100 billion kelvin, about ten thousand times hotter than the core of our Sun.

If a single star is less massive than about twenty-five times the mass of the Sun (but more than about ten times the mass of the Sun), the collapse creates a brand new kind of object at its center: a neutron star. Single stars with a mass between eight and about ten times the mass of the Sun also end up as neutron stars, but their nuclear evolution in the core (not discussed here) differs from the above scenario.

At the high density of the collapsing core, electrons and protons merge. An individual electron's negative charge cancels out a proton's positive charge, and they unite to create a neutron and a neutrino. Individual nuclei no longer exist; they have disappeared into a mass of what is known as degenerate neutron matter. (Finally, some juicy names!) I love the name of the countervailing pressure: neutron degeneracy pressure. If this would-be neutron star grows *more* massive than about 3 solar masses, which is the case if the single star's mass (called the progenitor) is larger than about twenty-five times the mass of the Sun, then gravity overpowers even the neutron degeneracy pressure, and what do you think will happen then? Take a guess.

That's right. I figured you guessed it. What else could it be but a black hole, a place where matter can no longer exist in any form we can understand; where, if you get close, gravity is so powerful that no radiation can escape: no light, no X-rays, no gamma rays, no neutrinos, no *anything*. The evolution in binary systems (see the next chapter) can be very different because in a binary the envelope of the massive star may be removed at an early stage, and the core mass may not be able to grow as much as in a single star. In that case even a star that originally was forty times more massive than the Sun may still leave a neutron star.

I'd like to stress that the dividing line between progenitors that form neutron stars and black holes is not clear cut; it depends on many variables other than just the mass of the progenitor; stellar rotation, for instance, is also important.

But black holes do exist—they aren't the invention of feverish scientists and science fiction writers—and they are incredibly fascinating. They are deeply involved in the X-ray universe, and I'll come back to them—I promise. For the moment, I'll just say this: not only are they real—they probably make up the nucleus of every reasonably massive galaxy in the universe.

Let's go back to the core collapse. Once the neutron star forms—remember, we're talking milliseconds here—the stellar matter still trying to race into it literally bounces off, forming an outward-going shock wave, which will eventually stall due to energy being consumed by the breaking apart of the remaining iron nuclei. (Remember that energy is released when light elements fuse to form an iron nucleus, therefore breaking an iron nucleus apart will consume energy.) When electrons and protons merge during core collapse to become neutrons, neutrinos are also produced. In addition, at the high core temperature of about 100 billion kelvin, so-called thermal neutrinos are produced. The neutrinos carry about 99 percent (which is about 10^{46} joules) of all energy released in the core collapse. The remaining 1 percent (10^{44} joules) is largely in the form of kinetic energy of the star's ejected matter.

The nearly massless and neutral neutrinos ordinarily sail through nearly all matter, and most do escape the core. However, because of the extremely high density of the surrounding matter, they transfer about 1 percent of their energy to the matter, which is then blasted away at speeds up to 20,000 kilometers per second. Some of this matter can be seen for thousands of years after the explosion—we call this a supernova remnant (like the Crab Nebula).

The supernova explosion is dazzling; the optical luminosity at maximum brightness is about 10^{35} joules per second. This is 300 million times the luminosity of our Sun, providing one of the great sights in the heavens when such a supernova occurs in our galaxy (which happens on average only about twice per century). Nowadays, with the use of fully automated robotic telescopes, many hundreds to a thousand supernovae are discovered each year in the large zoo of relatively nearby galaxies.

A core-collapse supernova emits two hundred times the energy that our Sun has produced in the past 5 billion years, and all that energy is released in roughly 1 second—and 99 percent comes out in neutrinos!

That's what happened in the year 1054, and the explosion produced the brightest star in our heavens in the past thousand years—so bright that it was visible in the daytime sky for weeks. A true cosmic flash in the interstellar pan, the supernova fades within a few years, as the gas cools and disperses. The gas doesn't disappear, though. That explosion in 1054 not only produced a solitary neutron star; it also produced the Crab Nebula, one of the more remarkable and still-changing objects in the entire sky, and a nearly endless source of new data, extraordinary images, and observational discoveries. Since so much astronomical activity takes place on an immense time scale, one we more often think of as geological—millions and billions of years—it's especially exciting when we find something that happens really fast, on a scale of seconds or minutes or even years. Parts of the Crab Nebula change shape every few days, and the Hubble Space Telescope and the Chandra X-Ray Observatory have found that the remnant of Supernova 1987A (located in the Large Magellanic Cloud) also changes shape in ways we can see.

Three different neutrino observatories on Earth picked up simultaneous neutrino bursts from Supernova 1987A, the light from which reached us on February 23, 1987. Neutrinos are so hard to detect that between them, these three instruments detected a total of just twenty-five in thirteen seconds, out of the roughly 300 trillion (3×10^{14}) neutrinos showering down in those thirteen seconds on every square meter of the Earth's surface directly facing the supernova. The supernova originally ejected something on the order of 10^{58} neutrinos, an almost unimaginably high number—but given its large distance from the Earth (about 170,000 light-years), "only" about 4×10^{28} neutrinos—thirty orders of magnitude fewer—actually reached the Earth. More than 99.9999999 percent go straight through the Earth; it would take a light-year (about 10^{13} kilometers) of lead to stop about half the neutrinos.

The progenitor of Supernova 1987A had thrown off a shell of gas about twenty thousand years earlier that had made rings around the star, and the rings remained invisible until about 8 months after the supernova explosion. The speed of the expelled gas was relatively slow—only around 8 kilometers per second—but over the years the shell's radius had reached a distance of about two-thirds of a light-year, about 8 light-months.

So the supernova went off, and about eight months later ultraviolet light from the explosion (traveling at the speed of light, of course) caught up with the ring of matter and turned it on, so to speak—and the ring started to emit visible light. You can see a picture of SN 1987A in the insert.

But there's more, and it involves X-rays. The gas expelled by the supernova in the explosion traveled at roughly 20,000 kilometers per second, only about fifteen times slower than the speed of light. Since we knew how far away the ring was by now, we could also predict when, approximately, the expelled matter was going to hit the ring, which it did a little over eleven years later, producing X-rays. Of course, we always have to remember that even though we talk about it as though it happened in the last few decades, in reality, since SN 1987A is in the Large Magellanic Cloud, it all happened about 170,000 years ago.

No neutron star has been detected to date in the remnant of SN 1987A. Some astrophysicists believe that a black hole was formed during core collapse after the initial formation of a neutron star. In 1990 I made a bet with Stan Woosley of the University of California, Santa Cruz; he is one of the world's experts on supernovae. We made a bet whether or not a neutron star would be found within five years. I lost the hundred-dollar bet.

There's more that these remarkable phenomena produce. In the superhot furnace of the supernova, higher orders of nuclear fusion slam nuclei together to create elements far heavier than iron that end up in gas clouds that may eventually coalesce and collapse into new stars and planets. We humans and all animals are made of elements that were cooked in stars. Without these stellar kilns, and without these stunningly violent explosions, the first of which was the big bang itself, we would never have the richness of elements that you see in the periodic table. So maybe we can think of a core-collapse supernova as resembling a celestial forest fire (a small one, to be sure), that in burning out one star creates the conditions for the birth of new stars and planets.

By any measure neutron stars are extreme objects. They are only a dozen miles across (smaller than some asteroids orbiting between Mars and Jupiter), about hundred thousand times smaller than the Sun, and thus about 300 billion (3×10^{14}) times more dense than the average density of the Sun. A teaspoon of neutron star matter would weigh 100 million tons on Earth.

One of the things I love about neutron stars is that simply saying or writing their name pulls together the two extremes of physics, the tiny and the immense, things so small we will never see them, in bodies so dense that they strain the capacity of our brains.

Neutron stars rotate, some of them at astonishing rates, especially when they first come into being. Why? For the same reason that an ice skater spinning around with her arms out spins more rapidly when she pulls them in. Physicists describe this by saying that angular momentum is conserved. Explaining angular momentum in detail is a bit complicated, but the idea is simple to grasp.

What does this have to do with neutron stars? Just this: Every object in the universe rotates. So the star that collapsed into the neutron star was rotating. It threw off most of its matter in the explosion but held on to one or two solar masses, now concentrated in an object a few thousand times smaller than the size of the core before collapse. Because angular momentum is conserved, neutron stars' rotational frequency therefore has to go up by at least a factor of a million.

The first two neutron stars discovered by Jocelyn Bell (see below) rotate about their axes in about 1.3 seconds. The neutron star in the Crab Nebula rotates about 30 times per second, while the fastest one that has been found so far rotates an astonishing 716 times per second! That means that the speed at the star's equator is about 15 percent of the speed of light!

The fact that all neutron stars rotate, and that many have substantial magnetic fields, gives rise to an important stellar phenomenon known as pulsars—short for "pulsating stars." Pulsars are neutron stars that emit beams of radio waves from their magnetic poles, which are, as in the case of the Earth, noticeably different from the geographic poles—the points at the end of the axis around which the star rotates. The pulsar's radio beam sweeps across the heavens as the star rotates. To an observer in the path of the beam, the star pulses at regular intervals, with the observer only seeing the beam for a brief moment. Astronomers sometimes call this the lighthouse effect, for obvious reasons. There are half a dozen known single neutron stars, not to be confused with neutron stars in binaries, which pulse over an extremely large range of the electromagnetic spectrum, including radio waves, visible light, X-rays, and gamma rays. The pulsar in the Crab Nebula is one of them.

Jocelyn Bell discovered the first pulsar in 1967 when she was a graduate student in Cambridge, England. She and her supervisor, Antony Hewish, at first didn't know what to make of the regularity of the pulsations, which lasted for only about 0.04 seconds and were about 1.3373 seconds apart (this is called the pulsar period). They initially called the pulsar LGM-1, for "Little Green Men," hinting that the regular pulsations might have been the product of extraterrestrial life. A second LGM was soon discovered by Bell with a period of about 1.2 seconds, and it became clear that the pulses were not produced by extraterrestrial life—why would two completely different civilizations send signals to Earth with about the same period? Shortly after Bell and Hewish published their results, it was recognized by Thomas Gold at Cornell University that pulsars were rotating neutron stars.

Black Holes

I told you we'd get here. It is finally time to look directly at these bizarre objects. I understand why people might be afraid of them—if you spend a little time on YouTube, you'll see dozens of "re-creations" of what black holes might look like, and most of them fall in the category of "death stars" or "star eaters." In the popular imagination black holes are superpowerful cosmic sinkholes, destined to suck everything into their insatiable maws.

But the notion that even a supermassive black hole swallows up everything in its vicinity is a complete fallacy. All kinds of objects, chiefly stars, will orbit a stellar mass black hole or even a supermassive black hole with great stability. Otherwise, our own Milky Way would have disappeared into the enormous 4-million-solar-mass black hole at its center.

So what do we know about these strange beasts? A neutron star can only contain up to about 3 solar masses before the gravitational pull collapses it to form a black hole. If the original single nuclear-burning star was more massive than about twenty-five times the mass of the Sun, at core collapse the matter would continue to collapse rather than stopping at the neutron star stage. The result? A black hole.

If black holes have companion stars in binary systems, we can measure their gravitational effect on their visible partners, and in some rare cases we can even determine their masses. (I talk about these systems in the next chapter.)

Instead of a surface, a black hole has what astronomers call an event horizon, the spatial boundary at which the black hole's gravitational power is so great that nothing, not even electromagnetic radiation, can escape the gravitational field. I realize this doesn't make much sense, so try to imagine that the black hole is like a heavy ball resting in the middle of a rubber sheet. It causes the center to sag, right? If you don't have a rubber sheet handy, try using an old stocking, or a pair of discarded pantyhose. Cut out as large a square as you can and put a stone in the middle. Then lift the square from the sides. You see immediately that the stone creates a funnel-like depression resembling a tornado spout. Well, you've just created a three-dimensional version of what happens in spacetime in four dimensions. Physicists call the depression a gravity well because it mimics the effect gravity has on spacetime. If you replace the stone with a larger rock, you'll have made a deeper well, suggesting that a more massive object distorts spacetime even more.

Because we can only think in three spatial dimensions, we can't really visualize what it would mean for a massive star to make a funnel out of four-dimensional spacetime. It was Albert Einstein who taught us to think about gravity in this way, as the curvature of spacetime. Einstein converted gravity into a matter of geometry, though not the geometry you learned in high school.

The pantyhose experiment is not ideal—I'm sure that will come as a relief to many of you—for a number of reasons, but the main one is that you can't really imagine a marble in a stable orbit around a rock-generated gravity well. In real astronomical life, however, many objects achieve stable orbits around massive bodies for many millions, even billions of years. Think of our Moon orbiting the Earth, the Earth orbiting the Sun, and the Sun and another 100 billion stars orbiting in our own galaxy.

On the other hand, the demonstration does help us visualize a black hole. We can, for instance, see that the more massive the object, the deeper the well and the steeper the sides, and thus the more energy it takes to climb out of the well. Even electromagnetic radiation escaping from the gravity of a massive star has its energy reduced, which means its frequency decreases and its wavelengths become longer. You already know that we call a shift to the less energetic end of the electromagnetic spectrum a redshift. In the case of a compact star (massive and small), there is a redshift caused by gravity, which we call a gravitational redshift (which should not be confused with redshift due to Doppler shift—see chapter 2 and the next chapter).

To escape from the surface of a planet or star, you need a minimum speed to make sure that you never fall back. We call this the escape velocity, which is about 11 kilometers per second (about 25,000 miles per hour) for the Earth. Therefore, all satellites bound to Earth can never have a speed larger than 11 kilometers per second. The higher the escape velocity, the higher the energy needed to escape, since this depends both on the escape velocity and on the mass, m, of the objects that want to escape (the required kinetic energy is $\frac{1}{2} mv^2$).

Perhaps you can imagine that if the gravity well becomes very, very deep, the escape velocity from the bottom of the well could become greater than the speed of light. Since this is not possible, it means that nothing can escape that very deep gravity well, not even electromagnetic radiation.

A physicist named Karl Schwarzschild solved Einstein's equations of general relativity and calculated what the radius of a sphere with a given mass would be that would create a well so deep that nothing could escape it—a black hole. That radius is known as the Schwarzschild radius, and its size depends on the mass of the object. This is the radius of what we call the event horizon.

The equation itself is breathtakingly simple, but it is only valid for nonrotating black holes, often referred to as Schwarzschild black holes.* The equation involves well-known constants and the radius works out to just a little bit less than 3 kilometers per solar mass. That's how we can calculate the size—that is to say, the radius of the event horizon—of a black hole of, for example, 10 solar masses is about 30 kilometers. We could also calculate the radius of the event horizon of a black hole with the mass of the Earth—it would be a little less than 1 centimeter—but there's no evidence that such black holes exist. So if the mass of our Sun were concentrated into a sphere about 6 kilometers across, would it be like a neutron star? No—under the gravitational attraction of that much mass packed into such a small sphere, the Sun's matter would have collapsed into a black hole.

Long before Einstein, in 1748, the English philosopher and geologist John Michell showed that there could be stars whose gravitational pull is so great that light could not escape. He used simple Newtonian mechanics (any of my freshmen can do this now in thirty seconds) and he ended up with the same result as Schwarzschild: if a star has a mass N times the mass of our Sun, and if its radius is less than $3N$ kilometers, light cannot escape. It is a remarkable coincidence that Einstein's theory of general relativity gives the same result as a simple Newtonian approach.

At the center of the spherical event horizon lies what physicists call

---

*For rotating black holes the event horizon is oblate—fatter at the equator—not spherical.

a singularity, a point with zero volume and infinite density, something bizarre that only represents the solution to equations, not anything we can grasp. What a singularity is really like, no one has any idea, despite some fantasizing. There is no physics (yet) that can handle singularities.

All over the web you can see animated videos of black holes, most of them at once beautiful and menacing, but nearly all immense beyond belief, hinting at destruction on a cosmic scale. So when journalists began writing about the possibility that the world's largest accelerator, CERN's Large Hadron Collider (LHC), near Geneva, might be able to create a black hole, they managed to stir up a good deal of concern among nonscientists that these physicists were rolling dice with the future of the planet.

But were they really? Suppose they *had* accidentally created a black hole—would it have started eating up the Earth? We can figure this out fairly easily. The energy level at which opposing proton beams collided in the LHC on March 30, 2010, was 7 teraelectron volts (TeV), 7 trillion electron volts, 3.5 trillion per beam. Ultimately, the LHC scientists plan to reach collisions of 14 TeV, far beyond anything possible today. The mass of a proton is about 1.6×10^{-24} grams. Physicists often say that the mass, m, of a proton is about 1 billion electron volts, 1 GeV. Of course, GeV is energy and not mass, but since $E = mc^2$ (c being the speed of light), E is often referred to as "the mass." On the Massachusetts Turnpike there are signs: "Call 511 for Travel Information." Every time I see one I think about electrons, as an electron's mass is 511 keV.

Assuming that all the energy of the 14 TeV collision went into creating a black hole, it would have a mass of about 14,000 times that of a proton, or about 2×10^{-20} grams. Boatloads of physicists and review committees evaluated a mountain of literature on the question, published their results, and concluded that there was simply nothing to worry about. You want to know why, right? Fair enough. OK, here's how the arguments go.

First, scenarios in which the LHC would have enough energy to create such tiny black holes (known as micro black holes) depend on the theory of something called large extra dimensions, which remains highly speculative, to say the least. The theory goes well beyond anything that's been experimentally confirmed. So the likelihood even of creating micro black holes is, to begin with, exceptionally slim.

Clearly, the concern would be that these micro black holes would somehow be stable "accretors"—objects that could gather matter, pull it into themselves, and grow—and start gobbling up nearby matter and, eventually, the Earth. But if there were such things as stable micro black holes, they would already have been created by enormously energetic cosmic rays (which do exist) smacking into neutron stars and white dwarfs—where they would have taken up residence. And since white dwarfs and neutron stars appear stable on a time scale of hundreds of millions, if not billions of years, there don't seem to be any tiny black holes eating them up from within. In other words, stable micro black holes appear to pose zero threat.

On the other hand, without the theory of extra dimensions, black holes with a mass smaller than 2×10^{-5} grams (called the Planck mass) could not even be created. That is to say, there is no physics (yet) that can deal with black holes of such small mass; we would need a theory of quantum gravity, which doesn't exist. Thus the question of what the Schwarzschild radius would be for a 2×10^{-20} gram micro black hole is also meaningless.

Stephen Hawking has shown that black holes can evaporate. The lower the mass of a black hole, the faster it will evaporate. A black hole of 30 solar masses would evaporate in about 10^{71} years. A supermassive black hole of 1 billion solar masses would last about 10^{93} years! So you may ask, how long would it take for a micro black hole of mass 2×10^{-20} grams to evaporate? It's an excellent question, but no one knows the answer—Hawking's theory does not work in the domain of black hole masses lower than the Planck mass. But, just for curiosity's sake, the lifetime of a black hole of 2×10^{-5} grams is about 10^{-39} seconds. So it seems that they evaporate faster than the time it takes to produce them. In other words, they cannot even be produced.

It clearly seems unnecessary to worry about possible 2×10^{-20} gram LHC micro black holes.

I realize that this didn't stop people from suing to prevent the LHC from starting operations. It makes me worry, however, about the distance between scientists and the rest of humanity and what a lousy job we scientists have done of explaining what we do. Even when some of the best physicists in the world studied the issue and explained why it wouldn't pose any problems, journalists and politicians invented scenarios and fanned public fears on the basis of almost nothing. Science fiction at some level appears more powerful than science.

There's nothing more bizarre than a black hole, I think. At least a neutron star makes itself known by its surface. A neutron star says, in a way, "Here I am, and I can show you that I have a surface." A black hole has no surface and emits nothing at all (apart from Hawking radiation, which has never been observed).

Why some black holes, surrounded by a flattish ring of matter known as an accretion disk (see the next chapter), shoot out extremely high energy jets of particles perpendicular to the plane of the accretion disk, though not from inside the event horizon, is one of the great unsolved mysteries. Take a look at this image: www.wired.com/wired science/2009/01/spectacular-new/.

Everything about the interior of a black hole, inside the event horizon, we have to derive mathematically. After all, nothing can come out, so we receive no information from inside the black hole—what some physicists with a sense of humor call "cosmic censorship." The black hole is hidden inside its own cave. Once you fall through the event horizon, you can never get out—you can't even send a signal out. If you've fallen through the event horizon of a supermassive black hole, you wouldn't even know that you've passed the event horizon. It doesn't have a ditch, or a wall, or a ledge you need to walk over. Nothing in your local environment changes abruptly when you cross the horizon. Despite all the relativistic physics involved, if you are looking at your wristwatch you wouldn't see it stop, or appear to go faster or slower.

For someone watching you from a distance, the situation is very different. What they see is not you; their eyes are receiving the *images* of you carried by light that leaves your body and climbs its way out of the black hole's gravity well. As you get closer and closer to the horizon, the well gets deeper and deeper. Light has to expend more of its energy climbing out of the well, and experiences more and more gravitational redshift. All emitted electromagnetic radiation shifts to longer and longer wavelengths (lower frequencies). You would look redder and redder, and then you would disappear as your emissions would move into longer and longer wavelengths, such as infrared light and then longer and longer radio waves and all wavelengths would become infinity as you cross the event horizon. So even before you crossed the threshold, to the distant observer you would have effectively disappeared.

The distant observer also measures a really unanticipated thing: light travels slower when it comes from a region close to the black hole! Now, this does not violate any postulates of relativity: local observers near the black hole always measure light traveling at the same speed c (about 186,000 miles per second). But distant observers measure the speed of light to be less than c. The images of you carried by the light you emitted toward your distant observer take longer to get to her than they would if you were not near a black hole. This has a very interesting consequence: the observer sees you slow down as you approach the horizon! In fact, the images of you are taking longer and longer to get to her, so everything about you seems in slow motion. To an observer on Earth, your speed, your movements, your watch, even your heartbeat slows down as you approach the event horizon, and will *stop* completely by the time you reach it. If it weren't for the fact that the light you emit near the horizon becomes invisible due to the gravitational redshift, an observer would see you "frozen" on the horizon's surface for all eternity.

For simplicity I have been ignoring the Doppler shift, which will be enormous because of your ever-increasing speed as you approach the event horizon. In fact, as you cross the event horizon, you will be moving with the speed of light. (For an observer on Earth, the effects of this Doppler shift will be similar to the effects of the gravitational redshift.)

After you have crossed the event horizon, when you can no longer communicate with the outside world, you will still be able to see out. Light coming from outside the event horizon would be gravitationally shifted to higher frequency and shorter wavelength, so you would see a blueshifted universe. (That would also be the case if you could stand on the surface of a neutron star as well, for the same reason.) However, since you are falling in at great speed, the outside world will be moving away from you, and thus the outside world will become redshifted as well (as a result of the Doppler effect). So what will be the result? Will the blueshift win or will the redshift win? Or will neither win?

I asked Andrew Hamilton at the University of Colorado, JILA, who is a world authority on black holes and, as I expected, the answer is not so simple. The blueshift and redshift more or less cancel for a free faller, but the outside world looks redshifted above, redshifted below, and blueshifted in horizontal directions. (You may enjoy looking at his "Journey into a Schwarzschild black hole" movies to see what it's like to be an object falling into a black hole: http://jila.colorado.edu/~ajsh/insidebh/schw.html.)

There wouldn't be anyplace to stand, however, since there's no surface. All the matter that created the black hole has collapsed into a point, a singularity. What about the tidal forces—wouldn't you be torn to bits by the fact that there will be a difference between the gravitational force on your head and your toes? (It's the same effect as the side of the Earth facing the Moon experiencing a larger attractive force than the side of the Earth that is farther away from the Moon; this causes tides on Earth.)

Indeed, you would be torn to bits; a Schwarzschild black hole of 3 solar masses would rip you apart 0.15 seconds before you crossed the event horizon. This phenomenon is very graphically called spaghettification and involves your body being stretched beyond imagining. Once you have crossed the event horizon, the various pieces of your body will reach the singularity in about 0.00001 seconds, at which time you will be crushed into a point of infinite density. For a 4-million-solar-mass black hole, like the one at the center of our galaxy, you would safely cross the event horizon without having any problems at all, at least at first, but sooner or later you will be shredded spaghetti style! (Believe me, it will be "sooner," because you have only about 13 seconds left before that happens and then, 0.15 seconds later, you will reach the singularity.)

The whole idea of black holes is truly bizarre for everyone, but especially for the many astrophysicists who observe them (such as my former graduate students Jeffrey McClintock and Jon Miller). We know that stellar-mass black holes exist. They were discovered in 1971 when optical astronomers demonstrated that Cyg X-1 is a binary star system and that one of the two stars is a black hole! I will tell you all about this in the next chapter. Ready?

CHAPTER 13

Celestial Ballet

I t will come as no surprise to you by now that many of the stars you see in the heavens, with or without a telescope of any kind, are a lot more complicated than distant versions of our own familiar Sun. You may not know that about a third of what you see aren't even single stars at all, but rather what we call binaries: pairs of stars that are gravitationally bound together, orbiting each other. In other words, when you look up at the night sky about a third of the stars you see are binary systems—even though they appear to you as a single star. There are even triple star systems—three stars orbiting one another—out there as well, though they are not nearly as common. Because many of the bright X-ray sources in our galaxy turned out to be binary systems, I had many dealings with them. They are fascinating.

Each star in a binary system travels around what we call the center of mass of the binary, a point located between the two stars. If the two stars have equal mass, then the center of mass is at equal distance from the center of both stars. If the masses are not the same, then the center of mass is closer to the more massive star. Since both complete an orbit in exactly the same amount of time, the more massive star must have a lower orbital speed than the less massive one.

To visualize this principle, imagine a dumbbell with a bar connecting two ends of equal mass, rotating around its midpoint. Now imagine a new dumbbell, 2 pounds on one end, 10 pounds on the other. The center of mass of this dumbbell is quite close to the heavier end, so when it rotates you can see that the larger mass has a smaller orbit, and that the smaller mass has farther to go in the same time. If these are stars instead of weighted ends, you can see that the lower-mass star zooms around its orbit at five times the speed of its larger, clunkier companion.

If one of the stars is much more massive than its companion, the center of mass of the system can even lie within the more massive star. In the case of the Earth and Moon (which is a binary system), the center of mass is about 1,700 kilometers (a little more than a thousand miles) below the Earth's surface. (I mention this in appendix 2.)

Sirius, the brightest star in the sky (at a distance from us of about 8.6 light-years), is a binary system made up of two stars known as Sirius A and Sirius B. They orbit their common center of mass about once every fifty years (we call this the orbital period).

How can we tell that we're looking at a binary system? We can't see binaries separately with the naked eye. Depending on the distance of the system and the power of the telescopes we're using, we can sometimes get visual confirmation by seeing the two stars as separate.

The famous German mathematician and astronomer Friedrich Wilhelm Bessel predicted that the brightest star in the sky, Sirius, was a binary system, consisting of a visible and an invisible star. He had concluded this based on his precise astronomical observations—he was the first in 1838 to make parallax observations (he narrowly beat Henderson—see chapter 2). In 1844 he wrote a famous letter to Alexander von Humboldt: "I adhere to the conviction that the star Sirius is a binary system consisting of a visible and an invisible star. There is no reason to suppose that luminosity is an essential quality of cosmic bodies. Visibility of countless stars is no argument against the invisibility of countless others." This is a statement of profound depth; what we can't see, we usually don't believe. Bessel started what we now call the astronomy of the invisible.

No one actually *saw* the "invisible" companion (called Sirius B) until 1862, when Alvan Clark was testing a brand new 18.5-inch telescope (the largest one at the time, made by his father's company) in my hometown, Cambridge, Massachusetts. He turned the telescope on Sirius as it was rising above the Boston skyline, for a test, and discovered Sirius B (it was ten thousand times fainter than Sirius A).

Thank Goodness for Stellar Spectroscopy: Blueshifts and Redshifts

By far the most common method of figuring out that stars are binaries, especially if they're distant, is by using spectroscopy and measuring what's known as the Doppler shift. There may be no more powerful astrophysical tool than spectroscopy, and no more important discovery in astronomy in the past several centuries than the Doppler shift.

You already know that when objects are hot enough they will emit visible light (blackbody radiation). By decomposing sunlight in the way a prism does, the raindrops that make up a rainbow (chapter 5) show you a continuum of colors from red at one end to violet at the other, called a spectrum. If you decompose the light from a star, you will also see a spectrum, but it may not have all the colors in equal strengths. The cooler the star, for example, the redder the star (and its spectrum) will be. The temperature of Betelgeuse (in the constellation Orion) is only 2,000 kelvin; it's among the reddest stars in the sky. The temperature of Bellatrix, on the other hand, also in Orion, is 28,000 kelvin; it's among the bluest and brightest stars in the sky and is often called the Amazon Star.

A close look at stellar spectra shows narrow gaps where colors are reduced or even completely absent, which we call absorption lines. The spectrum of the Sun shows thousands of such absorption lines.

These are caused by the many different elements in the atmospheres of the stars. Atoms, as you know, are made of nuclei and electrons. The electrons cannot just have any energy; they have discrete energy levels—they cannot have energies in between these distinct levels. Their energies, in other words, are "quantized"—the term that gives rise to the field of quantum mechanics.

Neutral hydrogen has one electron. If it is bombarded with light, this electron can jump from one energy level to a higher energy level by absorbing the energy of a light photon. But because of the quantization of the energy levels of the electron, this cannot happen with photons of just any energy. Only those photons that have just the right energy (thus exactly the right frequency and wavelength) for the electron to make the quantum jump from one level to another will do. This process (called resonance absorption) kills these photons and creates an absence at that frequency in the continuum spectrum, which we call an absorption line.

Hydrogen can produce four absorption lines (at precisely known wavelengths, or colors) in the visible part of a stellar spectrum. Most elements can produce many more lines, because they have lots more electrons than hydrogen. In fact, each element has its own unique combination of absorption lines, which amounts to a fingerprint. We know these very well from studying and measuring them in the laboratory. A careful study of the absorption lines in a stellar spectrum can therefore tell us which elements are present in the star's atmosphere.

However, when a star moves away from us, the phenomenon known as the Doppler shift causes the star's entire spectrum (including the absorption lines) to shift toward the red part of the spectrum (we call this redshift). If, by contrast, the spectrum is blueshifted, we know the star is moving toward us. By carefully measuring the amount of shift in the wavelength of a star's absorption lines, we can calculate the speed with which the star is moving relative to us.

If we observe a binary system, for example, each star will move toward us for half of its orbit and away from us during the other half. Its companion will be doing exactly the opposite. If both stars are bright enough, we will see redshifted *and* blueshifted absorption lines in the spectrum. That would tell us that we are looking at a binary system. But the absorption lines will be moving along the spectrum due to the orbital motion of the stars. As an example, if the orbital period is twenty years, each absorption line will make a complete excursion in twenty years (ten years of redshift and ten years of blueshift).

If we can see only redshifted (or only blueshifted) absorption lines, we still know it is a binary system if we see the absorption lines move back and forth in the spectrum; a measurement of the time it takes for a full cycle of the lines will tell us the orbital period of the star. When would this happen? In the event that one star is too faint to be seen from Earth in optical light.

Let's now return to our X-ray sources.

Shklovsky and Beyond

Way back in 1967, the Russian physicist Joseph Shklovsky had proposed a model for Sco X-1. "By all its characteristics, this model corresponds to a neutron star in a state of accretion . . . the natural and very efficient supply of gas for such an accretion is a stream of gas which flows from a secondary component of a close binary system toward the primary component which is a neutron star."

I realize these lines may not strike you as earthshaking. It doesn't help that they are written in the rather dry technical language of astrophysics. But that's the way professionals in just about any field talk to one another. My purpose in the classroom, and the main reason I've written this book, is to translate the truly astounding, groundbreaking, sometimes even revolutionary discoveries of my fellow physicists into concepts and language intelligent, curious laypeople can really get hold of—to make a bridge between the world of professional scientists and your world. Too many of us seem to prefer talking only to our peers and make it awfully difficult for most people—even those who really want to understand science—to enter our world.

So let's take Shklovsky's idea and see what he was proposing: a binary star system composed of a neutron star and a companion from which matter was flowing to the neutron star. The neutron star would then be "in a state of accretion"—in other words, it would be accreting matter from its companion, the donor star. What a bizarre idea, right?

Shklovsky turned out to be right. But here's the funny thing. He was only talking about Sco X-1 at the time, and most of us didn't take his idea too seriously. But that's often the case with theory. I don't think I would be offending any of my theoretician colleagues by saying that the great majority of theory in astrophysics turns out to be wrong. So of course many of us in observational astrophysics don't pay much attention to most of it.

It turns out that accreting neutron stars are in fact the perfect environments to produce X-rays. How did we find out that Shklovsky was right?

It took until the early seventies for astronomers to nail down the binary nature of some X-ray sources—but that didn't necessarily mean that they were accreting neutron stars. The very first source to reveal its secrets was Cyg X-1, and it turned out to be one of the most important in all of X-ray astronomy. Discovered during a rocket flight in 1964, it is a very bright and powerful source of X-rays, so it has attracted the attention of X-ray astronomers ever since.

Radio astronomers then discovered radio waves from Cyg X-1 in 1971. Their radio telescopes pinpointed Cyg X-1's position to a region (an error box) in the sky of about 350 square arc seconds, about twenty times smaller than had been possible by tracking its X-rays. They went looking for its optical counterpart. In other words, they wanted to *see*, in visible light, the star that was generating the mysterious X-rays.

There was a very bright blue supergiant known as HDE 226868 in the radio error box. Given the kind of star it was, astronomers could make comparisons with other very similar stars to make a pretty good estimate of its mass. Five astronomers, including the world-famous Allan Sandage, concluded that HDE 226868 was just a "normal B0 supergiant, with no peculiarities," and they dismissed the fact that it was the optical counterpart of Cyg X-1. Other (at the time less famous) optical astronomers examined the star more closely and made some earthshaking discoveries.

They discovered that the star was a member of a binary system with an orbital period of 5.6 days. They argued correctly that the strong X-ray flux from this binary system was due to the accretion of gas from the optical star (the donor) to a very small—compact—object. Only a gas flow onto a massive but very small object could explain the copious X-ray flux.

They made Doppler-shift measurements of absorption lines in the spectrum of the donor star as it moved around in its orbit (remember, as it moved toward Earth, the spectra would shift toward the blue end, and as it moved away, it would shift toward the red) and concluded that the X-ray-generating companion star was too massive to be either a neutron star or a white dwarf (another compact, very dense star, like Sirius B). Well, if it couldn't be either of those, and it was even more massive than a neutron star, what else could it be? Of course—a black hole! And that's what they proposed.

As observational scientists, however, they stated their conclusions more circumspectly. Louise Webster and Paul Murdin, whose discovery ran in *Nature* on January 7, 1972, put it this way: "The mass of the companion being probably larger than 2 solar masses, it is inevitable that we should also speculate that it might be a black hole." Here's what Tom Bolton wrote a month later in *Nature*: "This raises the distinct possibility that the secondary [the accretor] is a black hole." A picture of an artistic impression of Cyg X-1 can be seen in the insert.

So these wonderful astronomers, Webster and Murdin in England and Bolton in Toronto, shared the discovery of X-ray binaries *and* finding the first black hole in our galaxy. (Bolton was so proud, he had the license plate Cyg X-1 for a number of years.)

I've always thought it was odd that they never received a major prize for their absolutely phenomenal discovery. After all, they hit the field at its heart, and they were *first*! They nailed the first X-ray binary system. And they said that the accretor was probably a black hole. What a piece of work!

In 1975 none other than Stephen Hawking bet his friend, fellow theoretical physicist Kip Thorne, that Cyg X-1 wasn't a black hole at all—even though most astronomers thought it was by then. He eventually conceded the bet, fifteen years later, I think to his own delight, since so much of his work has revolved around black holes. The most recent (soon to be published) and most accurate measurement of the mass of the black hole in Cyg X-1 is about 15 solar masses (private communication from Jerry Orosz and my former student Jeff McClintock).

If you're sharp, I know you're already thinking, "Hold it! You just said black holes don't emit anything, that nothing can escape their gravitational field—how can they emit X-rays?" Terrific question, which I promise to answer eventually, but here's a preview: the X-rays emitted by a black hole do not come from inside the event horizon—they're emitted by matter on the way *into* the black hole. While a black hole explained our observations of Cyg X-1, it could not explain what was seen in terms of X-ray emission from other binary stars. For that we needed neutron star binaries, which were discovered with the wonderful satellite Uhuru.

The field of X-ray astronomy dramatically changed in December 1970, when the first satellite totally dedicated to X-ray astronomy went into orbit. Launched from Kenya on the seventh anniversary of Kenyan independence, the satellite was named Uhuru, Swahili for "freedom."

Uhuru began a revolution that hasn't stopped to this day. Think about what a satellite could do. Observations 365 days a year, twenty-four hours a day, with no atmosphere at all! Uhuru was able to observe in ways we could only have dreamed about a half dozen years earlier. In just a little over two years, Uhuru mapped the X-ray sky with counters that could pick up sources five hundred times fainter than the Crab Nebula, ten thousand times fainter than Sco X-1. It found 339 of them (we'd only found several dozen before that) and provided the first X-ray map of the entire sky.

Freeing us at last from atmospheric shackles, satellite observatories have reshaped our view of the universe, as we learned to see deep space—and the astonishing objects it contains—through every area of the electromagnetic spectrum. The Hubble Space Telescope expanded our view of the optical universe, while a series of X-ray observatories did the same for the X-ray universe. Gamma-ray observatories are now observing the universe at even higher energies.

In 1971 Uhuru discovered 4.84-second pulsations from Cen X-3 (in the constellation Centaurus). During a one-day interval Uhuru observed a change in the X-ray flux by a factor of ten in about one hour. The period of the pulsations first decreased and then increased by about 0.02 and 0.04 percent, each change of period occurring in about an hour. All this was very exciting but also very puzzling. The pulsations couldn't be the result of a spinning neutron star; their rotation periods were known to be steady like a rock. None of the known pulsars could possibly change their period by 0.04 percent in an hour.

The entire picture came together beautifully when the Uhuru group later discovered that Cen X-3 was a binary system with an orbital period of 2.09 days. The 4.84-second pulsations were due to the rotation of the accreting neutron star. The evidence was overwhelming. First, they clearly saw repetitive eclipses (every 2.09 days) when the neutron star hides behind the donor star, blocking the X-rays emissions. And second, they were able to measure the Doppler shift in the periods of the pulsations. When the neutron star is moving toward us, the pulsation period is a little shorter, a little longer when moving away. These earthshaking results were published in March 1972. All this naturally explained the phenomena that seemed so puzzling in the 1971 paper. It was just as Shklovsky had predicted for Sco X-1: a binary system with a donor star and an accreting neutron star.

Later that very same year, Giacconi's group found yet another source, Hercules X-1 (or Her X-1, as we like to say), with pulsations and eclipses. Another neutron star X-ray binary!

These were absolutely stunning discoveries that transformed X-ray astronomy, dominating the field for decades to come. X-ray binaries are very rare; perhaps only one in a hundred million binary stars in our galaxy is an X-ray binary. Even so, we now know that there are several hundred X-ray binaries in our galaxy. In most cases the compact object, the accretor, is a white dwarf or a neutron star, but there are at least two dozen known systems in which the accretor is a black hole.

Remember the 2.3-minute periodicity that my group discovered in 1970 (before the launch of Uhuru)? At the time we had no clue what these periodic changes meant. Well, we now know that GX 1+4 is an X-ray binary system with an orbital period of about 304 days, and the accreting neutron star spins around in about 2.3 minutes.

X-ray Binaries: How They Work

When a neutron star pairs up with the right-size donor star at the right distance, it can create some amazing fireworks. There, in the reaches of space, stars Isaac Newton could never even have imagined perform a beautiful dance, all the while utterly bound by the laws of classical mechanics any undergraduate science major can grasp.

To understand this better, let's start close to home. The Earth and the Moon are a binary system. If you draw a line from the center of the Earth to the center of the Moon, there is a point on that line where the gravitational force toward the Moon is equal but opposite to the gravitational force toward Earth. If you were there, the net force on you would be zero. If you were on one side of that point you would fall to Earth; if you were on the other side you would fall toward the Moon. That point has a name; we call it the inner Lagrangian point. Of course, it lies very close to the moon, because the Moon's mass is about eighty times smaller than that of the Earth.

Let's now return to X-ray binaries consisting of an accreting neutron star and a much larger donor star. If the two stars are very close to each other, the inner Lagrangian point can lie below the surface of the donor star. If that is the case, some of the matter of the donor star will experience a gravitational force toward the neutron star that is larger than the gravitational force toward the center of the donor star. Consequently matter—hot hydrogen gas—will flow from the donor star to the neutron star.

Since the stars are orbiting their common center of mass, the matter cannot fall directly toward the neutron star. Before it reaches the surface, the matter falls into an orbit around the neutron star, creating a spinning disk of hot gas that we call an accretion disk. Some of the gas on the inner ring of the disk ultimately finds its way down to the surface of the neutron star.

Now an interesting piece of physics gets involved that you are already familiar with in another context. Since the gas is very hot, it is ionized, consisting of positively charged protons and negatively charged electrons. But since the neutron stars have very strong magnetic fields, these charged particles are forced to follow the star's magnetic field lines, so most of this plasma ends up at the magnetic poles of the neutron star, like the aurora borealis on Earth. The neutron star's magnetic poles (where matter crashes onto the neutron star) become hot spots with temperatures of millions of degrees kelvin, emitting X-rays. And as magnetic poles generally do not coincide with the poles of the axis of rotation (see chapter 12), we on Earth will only receive a high X-ray flux when a hot spot is facing us. Since the neutron star rotates, it appears to pulsate.

Every X-ray binary has an accretion disk orbiting the accretor, be it a neutron star, a white dwarf or, as in Cyg X-1, a black hole. Accretion disks are among the most extraordinary objects in the universe, and almost no one except professional astronomers has ever even heard of them.

There are accretion disks around all black hole X-ray binaries. There are even accretion disks orbiting supermassive black holes at the center of many galaxies, though probably not, as it turns out, around the supermassive black hole at the center of our own galaxy.

The study of accretion disks is now an entire field of astrophysics. You can see some wonderful images of them here: www.google.com/images?hl=en&q=xray+binaries&um=1&ie=UTF. There is still lots about accretion disks that we don't know. One of the most embarrassing problems is that we still don't understand well how the matter in the accretion disks finds its way to the compact object. Another remaining problem is our lack of understanding of instabilities in the accretion disks, which give rise to variability in the matter flow onto the compact object, and the variability in X-ray luminosity. Our understanding of radio jets present in several X-ray binaries is also very poor.

A donor star can transfer up to about 10^{18} grams per second to the accreting neutron star. It sounds like a lot, but even at that rate it would take two hundred years to transfer an amount of matter equal to the Earth's mass. Matter from the disk flows toward the accretor in the grip of its intense gravitational field, which accelerates the gas to an extremely high speed: about one third to one half the speed of light. Gravitational potential energy released by this matter is converted into kinetic energy (roughly 5×10^{30} watts) and heats the racing hydrogen gas to a temperature of millions of degrees.

You know that when matter is heated it gives off blackbody radiation (see chapter 14). The higher the temperature, the more energetic the radiation, making shorter wavelengths and higher frequencies. When matter reaches 10 to 100 million kelvin, the radiation it generates is mostly in X-rays. Almost all 5×10^{30} watts are released in the form of X-rays; compare that with the total luminosity of our Sun (4×10^{26} watts) of which only about 10^{20} watts is in the form of X-rays. Our Sun's surface temperature is a veritable ice cube in comparison.

The neutron stars themselves are much too small to be seen optically—but we can see the much larger donor stars and the accretion disks with optical telescopes. The disks themselves can radiate quite a bit of light partly due to a process called X-ray heating. When the matter from the disk crashes onto the surface of the neutron star, the resultant X-rays go off in all directions and thus also slam into the disk itself, heating it to even higher temperatures. I will tell you more about that in the next chapter, on X-ray bursts.

The discovery of X-ray binaries solved the first mystery of extrasolar X-rays. We now understand why the X-ray luminosity of a source like Sco X-1 is ten thousand times greater than its optical luminosity. The X-rays come from the very hot neutron star (with temperatures of tens of millions kelvin), and the optical light comes from the much cooler donor star and the accretion disk.

When we thought that we had a fair understanding of how X-ray binaries work, nature had another surprise in store for us. The X-ray astronomers began making observational discoveries that were outstripping the theoretical models.

In 1975, the discovery of something truly bizarre led to a high point of my scientific career. I became completely immersed in the effort to observe, study, and explain these remarkable and mysterious phenomena: X-ray bursts.

Part of the story about X-ray bursts includes a battle I had with Russian scientists who completely misinterpreted their data and also with some of my colleagues at Harvard who believed that X-ray bursts were produced by very massive black holes (poor black holes, they have been unjustly blamed for so much). Believe it or not, I was even called (more than once) to not publish some data on bursts for reasons of national security.

CHAPTER 14

X-ray Bursters!

Nature is always full of surprises, and in 1975 it rocked the X-ray community. Things became so intense that emotions at times got out of control, and I was in the middle of it all. For years I was arguing with a colleague of mine at Harvard (who would not listen), but I had more luck with my Russian colleagues (who did listen). Because of my central role in all of this it may be very difficult for me to be objective, but I'll try!

The new thing was X-ray bursts. They were discovered independently in 1975 by Grindlay and Heise using data from the Astronomical Netherlands Satellite (ANS) and by Belian, Conner, and Evans, using data from the United States' two Vela-5 spy satellites designed to detect nuclear tests. X-ray bursts were a completely different animal from the variability we discovered from Sco X-1, which had a flare-up by a factor of four over a ten-minute period that lasted tens of minutes. X-ray bursts were much faster, much brighter, and they lasted only tens of seconds.

At MIT we had our own satellite (launched in May 1975) known as the Third Small Astronomy Satellite, or SAS-3. Its name wasn't as romantic as "Uhuru," but the work was the most absorbing of my entire life. We had heard about bursters and began looking for them in January 1976; by March we'd found five of our own. By the end of the year we'd found a total of ten. Because of the sensitivity of SAS-3, and the way it was configured, it turned out to be the ideal instrument to discover burst sources and to study them. Of course, it wasn't specially designed to detect X-ray bursts; so in a way it was a bit of luck. You see what a leading role Lady Luck has played in my life! We were getting amazing data—a bit of gold pouring out of the sky every day, twenty-four hours a day—and I worked around the clock. I was dedicated, but also obsessed. It was a once in a lifetime opportunity to have an X-ray observatory you can point in any direction you want to and get data of high quality.

The truth is that we all caught "burst fever"—undergraduates and graduate students, support staff and postdocs and faculty—and I can still remember the feeling, like a glow. We ended up in different observing groups, which meant that we got competitive, even with one another. Some of us didn't like that, but I have to say that I think it pushed us to do more and better, and the scientific results were just fantastic.

That level of obsession was not good for my marriage, and not good for my family life either. My scientific life was immeasurably enhanced, but my first marriage dissolved. Of course it was my fault. For years I'd been going away for months at a time to fly balloons halfway around the globe. Now that we had our own satellite, I might as well still have been in Australia.

The burst sources became a kind of substitute family. After all, we lived with them and slept with them and learned them inside out. Like friends, each one was unique, with its own idiosyncrasies. Even now, I recognize many of these telltale burst profiles.

Most of these sources were about 25,000 light-years away, which allowed us to calculate that the total X-ray energy in a burst (emitted in less than a minute) was about 10^{32} joules, a number that's almost impossible to grasp. So look at it this way: it takes our Sun about three days to emit 10^{32} joules of energy in all wavelengths.

Some of these bursts came with nearly clocklike regularity, such as the bursts from MXB 1659-29, which produced bursts at 2.4-hour intervals, while others changed their burst intervals from hours to days, and some showed no bursts at all for several months. The M in MXB stands for MIT, the X for X-rays, and the B for burster. The numbers indicate the source's celestial coordinates in what's known as the equatorial coordinate system. For the amateur astronomers among you, this will be familiar.

The key question, of course, was what caused these bursts? Two of my colleagues at Harvard (including Josh Grindlay, who was one of the codiscoverers of X-ray bursts) got carried away and proposed in 1976 that the bursts were produced by black holes with a mass greater than several hundred times the mass of the Sun.

We soon discovered that the spectra during X-ray bursts resemble the spectra from a cooling black body. A black body is not a black hole. It's an ideal construct to stand in for an object that absorbs all the radiation that strikes it, rather than reflecting any of it. (As you know, black absorbs radiation, while white reflects it—which is why in summer in Miami a black car left in a beach parking lot will always be hotter inside than a white one.) The other thing about an ideal black body is that since it reflects nothing, the only radiation it can emit is the result of its own temperature. Think about a heating element in an electric stove. When it reaches a cooking temperature, it begins to glow red, emitting low-frequency red light. As it gets hotter it reaches orange, then yellow, and usually not much more. When you turn off the electricity, the element cools, and the radiation it emits has a profile more or less like the tail end of bursts. The spectra of black bodies are so well known that if you measure the spectrum over time, you can calculate the temperature as it cools.

Since black bodies are very well understood, we can deduce a great deal about bursts based on elementary physics, which is quite amazing. Here we were, analyzing X-ray emission spectra of unknown sources 25,000 light-years away, and we made breakthroughs using the same physics that first-year college students learn at MIT!

We know that the total luminosity of a black body (how much energy per second it radiates) is proportional to the fourth power of its temperature (this is by no means intuitive), and it is proportional to its surface area (that's intuitive—the larger the area, the more energy can get out). So, if we have two spheres a meter in diameter, and one is twice as hot as the other, the hotter one will emit sixteen times (2^4) more energy per second. Since the surface area of a sphere is proportional to the square of its radius, we also know that if an object's temperature stays the same but triples in size, it will emit nine times more energy per second.

The X-ray spectrum at any moment in time of the burst tells us the blackbody temperature of the emitting object. During a burst, the temperature quickly rises to about 30 million kelvin and decreases slowly thereafter. But since we knew the approximate distance to these bursters, we could also calculate the luminosity of the source at any moment during the burst. But once you know both the blackbody temperature and the luminosity, you can calculate the radius of the emitting object, and that too can be done for any moment in time during the burst. The person who did this first was Jean Swank of NASA's Goddard Space Flight Center; we at MIT followed quickly and concluded that the bursts came from a cooling object with a radius of about 10 kilometers. This was strong evidence that the burst sources were neutron stars, not very massive black holes. And if they were neutron stars, they were probably X-ray binaries.

The Italian astronomer Laura Maraschi was visiting MIT in 1976, and one day in February she walked into my office and suggested that the bursts were the result of thermonuclear flashes, huge thermonuclear explosions on the surface of accreting neutron stars. When hydrogen accretes onto a neutron star, gravitational potential energy is converted to such tremendous heat that X-rays are emitted (see previous chapter). But as this accreted matter accumulates on the surface of the neutron star, Maraschi suggested, it might undergo nuclear fusion in a runaway process (like in a hydrogen bomb) and that might cause an X-ray burst. The next explosion might go off a few hours later when enough new nuclear fuel had been accreted to ignite. Maraschi demonstrated with a simple calculation on my blackboard that matter racing at roughly half the speed of light to the surface of a neutron star releases much more energy than what is released during the thermonuclear explosions, and that is what the data showed.

I was impressed—this explanation made sense to me. Thermonuclear explosions fit the bill. The cooling pattern we'd observed during the bursts also made sense if what we were seeing was a massive explosion on a neutron star. And her model explained the interval between bursts well since the amount of matter required for an explosion had to build up over time. At the normal rate of accretion, it should take a few hours to build up a critical mass, which was the kind of interval we found in many burst sources.

I keep a funny kind of radio in my office that always unsettles visitors. It's got a solar-powered battery inside, and it works only when the battery has enough juice. As the radio sits there soaking up sunlight, it slowly fills up with juice (a lot more slowly in the winter), then every ten minutes or so—sometimes longer if the weather's rotten—it suddenly starts playing, but only for a couple of seconds, as it quickly exhausts its supply of electricity. You see? The buildup in its battery is just like the buildup of accreted matter on the neutron star: when it gets to the right amount, the explosion goes off, and then fades away.

Then, several weeks after Maraschi's visit, on March 2, 1976, in the middle of burst fever, we discovered an X-ray source that I named MXB 1730-335 that was producing *a few thousand bursts per day*. The bursts came like machine-gun fire—many were only 6 seconds apart! I don't know if I can completely convey just how bizarre this seemed to us. This source (now called the Rapid Burster) was a complete outlier, and it immediately killed Maraschi's idea. First, there is no way that a sufficient amount of nuclear fuel could build up in six seconds on the surface of a neutron star to produce a thermonuclear explosion. Not only that, but if the bursts were a by-product of accretion, we should see a strong X-ray flux due to accretion alone (release of gravitational potential energy), far exceeding the energy present in the bursts, but that was not the case. So it seemed in early March 1976 that Maraschi's wonderful thermonuclear model for the bursts was as dead as the proverbial doornail. In our publication on MXB 1730-335, we suggested that the bursts are caused by "spasmodic accretion" onto a neutron star. In other words, what in most X-ray binaries is a steady flow of hot matter from the accretion disk onto the neutron star is very irregular in the case of the Rapid Burster.

When we measured the bursts over time, we found that the bigger the burst, the longer the wait before the next one. The waiting time to the next burst could be as short as six seconds and as long as eight minutes. Lightning does something similar. When there's a particularly large lightning bolt, the large discharge means that the wait needs to be longer for the electric field to build up its potential to the point that it can discharge again.

Later that year a translation of a 1975 Russian paper about X-ray bursts surfaced out of nowhere; it had been reporting burst detections made in 1971 with the Kosmos 428 satellite. We were stunned; the Russians had discovered X-ray bursts, and they had beaten the West! However, as I heard more and more about these bursts, I became very skeptical. Their bursts behaved so very, very differently from the many bursts that I had detected with SAS-3 that I began to seriously doubt whether the Russian bursts were real. I suspected that they were either man-made or produced near Earth in some odd, bizarre way. The iron curtain made it difficult to pursue this; there was no way to find out. However, I had the good fortune to be invited to attend a high-level conference in the Soviet Union in the summer of 1977. Only twelve Russians and twelve U.S. astrophysicists had been invited. That's where I met for the first time the world famous scientists Joseph Shklovsky, Roald Sagdeev, Yakov Zel'dovich, and Rashid Sunyaev.

I gave a talk on—you guessed it—X-ray bursts, and I got to meet the authors of the Russian burst paper. They generously showed me data of many bursts, way more than they had published in 1975. It was immediately obvious to me that all this was nonsense, but I did not tell them that, at least not at first. I first went to see their boss, Roald Sagdeev, who at the time was the director of the Space Research Institute of the USSR Academy of Sciences in Moscow. I told him that I wanted to discuss something rather delicate with him. He suggested we not do that in his office (bugs were all over the place), so we went outside. I gave him my reasons why their bursts were not what they thought they were—he immediately understood. I told him that I was afraid that my telling the world about this might get these guys into deep trouble under the Soviet regime. He assured me that that would not be the case, and he encouraged me to meet with them and tell them exactly what I had told him. So I did, and that was the last we ever heard of the Russian X-ray bursts. I'd like to add that we are still friends!

You may be curious to know what caused these Russian bursts. At the time I had no idea, but now I know; they were man-made, and guess who made them—the Russians! I'll solve this mystery in a bit.

For now let's return to the real X-ray bursts, which we were still trying to figure out. When the X-rays of the bursts plow into the accretion disk (or into the donor star) of an X-ray binary, the disk and the star get hotter and light up briefly in the optical part of the spectrum. Since the X-rays would first have to travel to the disk and donor star, we expected that any optical flash from the disk would reach us seconds after the X-ray burst. So we went hunting for coordinated X-ray and optical bursts. My former graduate student Jeff McClintock and his co-workers had made the first two optical identifications of burst sources (MXB 1636-53 and MXB 1735-44) in 1977. These two sources became our targets.

You see how science works? If a model is correct, then it ought to have observable consequences. In the summer of 1977 my colleague and friend Jeffrey Hoffman and I organized a worldwide simultaneous X-ray, radio, optical, and infrared "burst watch."

This was an amazing adventure all by itself. We had to convince astronomers at forty-four observatories in fourteen countries to devote precious observing time during the most favorable hours (known as "dark time," when the Moon is absent) staring at one faint star—that might do nothing. That they were willing to participate shows you just how significant astronomers considered the mystery of X-ray bursts. Over thirty-five days, with SAS-3, we detected 120 X-ray bursts from the burst source MXB 1636-53 but absolutely no bursts were observed with the telescopes on the ground. What a disappointment!

You might imagine that we had to apologize to our colleagues around the world, but the truth is that none saw it as a problem. This is what science is all about.

So we tried again the following year using only large ground-based telescopes. Jeff Hoffman had left for Houston to become an astronaut, but my graduate student Lynn Cominsky and the Dutch astronomer Jan van Paradijs (who had come to MIT in September 1977) joined me in the 1978 burst watch.* This time we selected MXB 1735–44. On the night of June 2, 1978, we succeeded! Josh Grindlay and his co-workers (including McClintock) detected an optical burst with the 1.5-meter telescope at Cerro Tololo in Chile a few seconds after we, at MIT, detected an X-ray burst with SAS-3. We made it to the front page of *Nature,* which was quite an honor. This work further supported our conviction that X-ray bursts come from X-ray binaries.

What was very puzzling to us was why all burst sources except one produce only a handful of bursts in a day and why the Rapid Burster was so very different. The answer lay with the most wonderful—and most bewildering—discovery of my career.

The Rapid Burster is what we call a transient. Cen X-2 is also a transient (see chapter 11). However, the Rapid Burster is what we call a recurrent transient. In the 1970s it became burst-active about every six months, but only for several weeks, after which it would go off the air.

About a year and a half after we discovered the Rapid Burster, we noticed something about its burst profiles that transformed this mystery source into a Rosetta Stone of X-ray bursters. In the fall of 1977, when the Rapid Burster was active again, my undergraduate student Herman Marshall looked very closely at the X-ray burst profiles and discovered a different kind of burst among the very rapid bursts, one that came far less frequently, about every three or four hours. These special bursts, as we called them at first, exhibited the same black body–like cooling pro-

---

*Little did I know at the time that Jan and I would become very close friends and that we would coauthor about 150 scientific publications before his untimely death in 1999.

file that characterized all the bursts from the many other burst sources. In other words, perhaps what we were calling special bursts—we soon called them Type I bursts, and gave the rapid bursts the designation Type II—weren't so special at all. The Type II bursts were clearly the result of spasmodic accretion—there was never any doubt about that—but maybe the common Type I bursts *were* due to thermonuclear flashes after all. I'll tell you shortly how we figured that out—just bear with me.

In the fall of 1978 my colleague Paul Joss at MIT had made some careful calculations about the nature of thermonuclear flashes on the surface of neutron stars. He concluded that the accumulated hydrogen first quietly fuses to helium, but that the helium, once it reaches a critical mass, pressure, and temperature, can then violently explode and produce a thermonuclear flash (thus a Type I burst). This led to a prediction that the X-ray energy released in the steady accretion should be roughly a hundred times larger than the energy released in the thermonuclear bursts. In other words, the available gravitational potential energy was roughly a hundred times larger than the available nuclear energy.

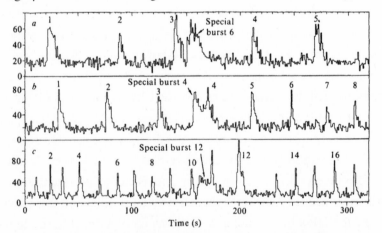

X-ray bursts from the Rapid Burster detected with SAS-3 in the fall of 1977. The height of the line represents the number of detected X-rays in about one second, while the horizontal axis represents time. Each panel shows about 300 seconds of data. The rapidly repetitive Type II bursts are numbered sequentially. One "Special Burst" is visible in each panel; they have different numbers. They are the Type I bursts (thermonuclear flashes). *This figure is from Hoffman, Marshall, and Lewin,* Nature, *16 Feb. 1978.*

We measured the total amount of energy emitted in X-rays from the Rapid Burster during the five-and-a-half days of our fall 1977 observations, and we found that about 120 times more energy was emitted in the Type II bursts than in the "special" Type I bursts. That was the clincher! At that point we knew that the Rapid Burster was an X-ray binary and that Type I bursts were the result of thermonuclear flashes on the surface of an accreting neutron star and that the Type II bursts were the result of the release of gravitational potential energy of the matter flowing from the donor star to the neutron star. There simply was no doubt about this anymore; from that time on, we knew that all Type I burst sources were neutron star X-ray binaries. At the same time we knew conclusively that black holes could not be the source of the bursts. Black holes have no surface, so they cannot produce thermonuclear flashes.

Even though it was already crystal clear to most of us by 1978 that burst sources were accreting neutron star binaries, Grindlay at Harvard continued to insist that the bursts were in fact produced by massive black holes. He even published a paper in 1978 in which he tried to explain how the bursts are produced by very massive black holes. I told you scientists can get emotionally attached to their theories. *The Real Paper* in Cambridge ran a long story, "Harvard and MIT at the Brink," featuring pictures of Grindlay and me.

Evidence for the binary nature of burst sources came in 1981 when my Danish friend Holger Pederson, Jan van Paradijs, and I discovered the 3.8-hour orbital period of the burst source MXB 1636–53. Yet, it was not until 1984 that Grindlay finally conceded.

So it was the weirdest X-ray source, the Rapid Burster, that helped confirm the theory of normal (Type I) X-ray bursts, which had been mystifying in their own right. The irony is that for all it explained, the Rapid Burster has remained mostly a mystery. Not so much for observers, but for theoreticians it remains an embarrassment. The best we could do, and in some ways the best we've ever been able to do, is come up with the explanation of "spasmodic accretion"—I know, it sounds like something you could catch on an exotic vacation. And the truth is, it's words, not physics. Somehow, the matter headed for the neutron star is temporarily held up in the disk before a blob or a ring of matter is released from the disk and spurts down to the surface of the star, releasing gravitational potential energy in bursts. We call this release a disk instability, but that too is just words; no one has a clue why and how it works.

Frankly, we also do not understand what the mechanism is behind the recurrent transient behavior of X-ray sources. Why do they turn on and off and on and off? We just don't know. Once in 1977 we started to pick up bursts simultaneously in all of SAS-3's detectors. This was bizarre, since they were viewing the sky in totally different directions. The only reasonable explanation we could come up with was that very-high-energy gamma rays were penetrating the entire spacecraft (something that X-rays cannot do) and leaving signals behind. Since all detectors "fired" at the same time, we had no clue what direction these gamma rays were coming from. After we had observed a few dozen of these episodes over a period of several months, they stopped. But thirteen months later they started up again. No one at MIT had a clue.

With the help of one of my undergraduate students, Christiane Tellefson, I started to catalog these bursts, and we even classified them as bursts A, B, and C, depending on their profiles. I stored them all in a file that I labeled SHIT BURSTS.

I remember giving a presentation to some people from NASA (who would visit us yearly), telling them our latest exciting news on X-ray bursts and showing them some of these bizarre bursts. I explained my reluctance to publish; they just didn't look kosher to me. However, they encouraged me not to delay publishing. So Christiane and I started to write a paper.

Then one day, completely out of the blue, I received a call from my former student Bob Scarlett, who was doing classified research at the Los Alamos National Laboratory. He asked me not to publish these weird bursts. I wanted an explanation, but he was not allowed to tell me why. He asked me to give him some of the times that these bursts had occurred, which I did. Two days later he called again and this time he *urged* me not to publish for reasons of national security. I nearly fell off my chair. I immediately called my friend France Córdova, who had once worked with me at MIT but who at that time was also working in Los Alamos. I told her about my conversations with Bob and hoped that she could cast some light on what was going on. She must have discussed it with Bob, because a few days later she too called and urged me not to publish. To put my mind at rest, she assured me that these bursts were of zero astronomical interest. To make a long story short, I did not publish.

Many years later I learned what had happened: the "shit bursts" had been produced by several Russian satellites that were powered by nuclear electrical generators, which contain extremely strong radioactive sources. Whenever SAS-3 came near any of the Russian satellites, they would shower our detectors with gamma rays emitted by the radioactive sources. Now, remember those weird bursts detected by the Russians back in 1971? I'm now quite certain these were also caused by the Russians' own satellites . . . what irony!

This period of my life, beginning in the late 1970s and going through 1995, was incredibly intense. X-ray astronomy was *the* cutting edge of observational astrophysics then. My involvement with X-ray bursts pushed me to the pinnacle of my scientific career. I probably gave a dozen colloquia yearly all over the world, in Eastern and Western Europe, Australia, Asia, Latin America, the Middle East, and throughout the United States. I got invited to give talks at many international astrophysics conferences and was the chief editor of three books on X-ray astronomy, the last one, *Compact Stellar X-ray Sources*, in 2006. It was a heady, wonderful time.

And yet, despite the amazing advances we made, the Rapid Burster has resisted all attempts to unlock its deepest mysteries. Someone will figure it out some day, I'm sure. And they in turn will be confronted with something equally perplexing. That's what I love about physics. And why I keep a poster-size reproduction of the Rapid Burster's burst profiles prominently displayed in my MIT office. Whether it's in the Large Hadron Collider or at the farthest reaches of the Hubble Ultra Deep Field, physicists are getting more and more data, and coming up with more and more ingenious theories. The one thing I know is whatever they find, and propose, and theorize, they'll uncover yet more mysteries. In physics, more answers lead to even more questions.

CHAPTER 15

Ways of Seeing

Most high school and college students hate taking physics because it is usually taught as a complicated set of mathematical formulas. That is not the approach I use at MIT, and it is not the approach I use in this book. I present physics as a way of seeing our world, revealing territories that would otherwise be hidden to us—from the tiniest subatomic particles to the vastness of our universe. Physics allows us to see the invisible forces at play all around us, from gravity to electromagnetism, and to be on the alert not only for where and when we'll find rainbows, but also halos, fogbows, and glories, and maybe even glassbows.

Each pioneering physicist changed the way we look at the world. After Newton, we could understand and predict the movements of the entire solar system, and we had the mathematics—calculus—to do so. After Newton, no one could claim that sunlight was not made up of colors, or that rainbows came from anything but sunlight refracting and reflecting in raindrops. After Maxwell, electricity and magnetism were forever linked: it was even hard for me to separate them into different chapters in this book.

This is why I see a fascinating relationship between physics and art; pioneering art is also a new way of seeing, a new way of looking at the world. You might be surprised to learn that for much of my life I've been almost as obsessed with modern art as I have been with physics; I have a love relationship with both! I've already mentioned my large collection of Fiestaware. I've also collected more than a hundred works of art—paintings, collages, sculptures, rugs, chairs, tables, puppets, masks—since the mid-sixties, and I no longer have enough wall or floor space in my home to display them all.

In my office at MIT, physics dominates, though I have two great works of art on loan from the university. But at home I probably only have about a dozen physics books—and about 250 art books. I was fortunate in being initiated into a love of art early.

My parents collected art, though they knew very little about it intellectually. They simply went by what they liked, which can lead down some blind alleys. Sometimes they picked some great works, and sometimes some not so great, or at least so it appears with the benefit of hindsight. One painting that made a strong impression on me is a portrait of my father, which I now have hanging over my fireplace in Cambridge. It is really very striking. My father was a real character—and like me, he was very opinionated. The artist, who knew him very well, caught him superbly, from the waist up, with his large, bald, oblong head sitting between his powerful square shoulders, his small mouth set in a self-satisfied smile. But it's his glasses that truly stand out: thick, black, outlining invisible eyes, they follow you around the room, while his left eyebrow arches quizzically over the frame. That was his entire personality: penetrating.

My father took me to art galleries and museums when I was in high school, and it was then that I really began to fall in love with art, as it taught me new ways of seeing. I loved that in galleries and museums, as opposed to school, you proceed according to your own interests, stopping when you wish, staying as long as you like, moving on when it suits you. You develop your own relationship to art. I soon started going to museums on my own, and before long, I had acquired a bit of knowledge. I plunged into van Gogh. (You know his name is really pronounced *van Chocch*—it's all but unpronounceable if you're not Dutch, two gutturals barely separated by a short O sound.) I ended up giving a lecture about van Gogh to my class when I was fifteen. I would also take my friends on tours to museums sometimes. So it was really art that got me into teaching.

This is when I first learned what a wonderful feeling it is to teach others—of any age—to expand their minds into new realms. It's a real shame that art can seem as obscure and difficult as so much of physics does to so many who had poor physics teachers. This is one reason that for the past eight years I've enjoyed putting an art quiz on my MIT bulletin board every week—an image I print off the web, with the question "Who is the artist?" I give prizes—some very nice art books—to the three contestants who have the most correct answers over the course of the year. Some regulars spend hours scouring the web and in doing so, they learn about art! I had so much fun with the weekly quiz that I've now put up a biweekly one on my Facebook page. You can try it yourself if you like.

I've also been lucky enough to have had some wonderful chances to collaborate with some amazing, cutting-edge artists in my life. In the late 1960s the German "sky artist" Otto Piene came to MIT as a fellow at the Center for Advanced Visual Studies, and later ended up directing it for two decades. Because I had already been flying some of my giant balloons by then, I got to help Otto make some of his sky art. The very first project we worked on together was called the *Light Line Experiment,* and consisted of four 250-foot-long polyethylene tubes filled with helium that, when held down at each end, made elegant arcs in the spring breezes at the MIT athletic fields. We tied all four together to make a thousand-foot-long balloon and let one end float up into the sky. At night we brought out spotlights that lit up parts of the snakelike balloons as they twisted and waved in the most amazing, constantly changing shapes, hundreds of feet in the air. It was fabulous!

My job in these projects was usually technical: figuring out whether Otto's ideas for the sizes and shapes of the balloons would be feasible. How thick should the polyethylene be, for example? We wanted it to be light enough to rise, but strong enough to stand up under windy conditions. At a 1974 event in Aspen, Colorado, we hung multifaceted glass beads from the tether lines of a "light tent." I made many calculations regarding the different balloon sizes and bead weights in order to get to a workable solution in terms of physics and aesthetics. I loved doing the physics to make Otto's artistic ideas a reality.

I got really involved with the immense, five-color *Rainbow* balloon he designed for the closing ceremonies of the 1972 Olympics in Munich. We of course had no idea that the Olympics would end so disastrously, with the massacre of the Israeli athletes, so our 1,500-foot *Rainbow,* which arched nearly five hundred feet high over the Olympic sea, became a symbol of hope in the face of catastrophe. A picture of the *Rainbow* balloon can be seen in the insert. When I began flying balloons to look at the universe, it never occurred to me that I could be involved in such projects.

Otto introduced me to the Dutch artist Peter Struycken, whose art I knew well because my parents had collected his works in the Netherlands. Otto called me up one day at MIT and said, "There's this Dutch artist in my office; would you like to meet him?" People always assume that if we're from the same little country we'd like to chat, but more often than not, I don't want to. I told Otto, "Why should I, what's his name?" When Otto said "Peter Struycken," of course I agreed, but in order to play it safe, I told Otto that I could only meet for half an hour (which was not true). So Peter came over to my office; we talked for almost five hours (yes, five hours!) and I invited him for oysters at Legal Sea Foods afterward! We clicked right from the start, and Peter became one of my closest friends for more than twenty years. This visit changed my life forever!

During that first discussion I was able to make Peter "see" why his major problem/question—"When is something different from something else?"—all depends on one's definition of difference. For some, a square may be different from a triangle and different from a circle. However, if you define geometric lines that close onto themselves as the same—well, then these three shapes are all the same.

Peter showed me a dozen computer drawings, all made with the same program, and he said, "They are all the same." To me they looked all very different. It all depends on one's definition of "the same." I added that if they were all the same to him, perhaps he would like to leave me one. He did and he wrote on it, in Dutch, *"Met dank voor een gesprek"* (literally, "With thanks for a discussion"). This was typical Peter: very very low key. Frankly, of the many Struyckens I have, this small drawing is my very favorite.

Peter had found in me a physicist who was not only very interested in art, but who could help him with his work. He is one of the world's pioneers in computer art. In 1979 Peter (with Lien and Daniel Dekkers) came for a year to MIT, and we started working together very closely. We met almost daily, and I had dinner at his place two or three times a week. Before Peter I "looked" at art—Peter made me "see" art.

Without him, I think I never would have learned to focus on pioneering works, to see how they can fundamentally transform our ways of seeing the world. I learned that art is not only, or even mostly, about beauty; it is about discovery, and this is where art and physics come together for me.

From that time on, I began to look at art very differently. What I "liked" was no longer important to me. What counted was the artistic quality, the new way of looking at the world, and that can only be appreciated if you really know something about art. I began to look closely at the years that works were made. Malevich's pioneering works of art from 1915 to 1920 are fascinating. Similar paintings made by others in the 1930s are of no interest to me. "Art is either plagiarism or revolution," said Paul Gauguin, with typical Gauguin arrogance, but there is some truth in it.

I was fascinated by the evolution that led to pioneering works. As an example, soon I was able to accurately tell the year that a Mondrian was made—his development between 1900 and 1925 was staggering—and my daughter Pauline can do that now too. Over the years I have noticed more than once that museums sometimes list the wrong date for a painting. When I point this out (as I always do), curators are sometimes embarrassed, but they always change it.

I worked with Peter on a dozen of his ideas. Our first project was "16th Space," art in sixteen dimensions (we beat string theory with its eleven dimensions). I also recall Peter's *Shift* series. He had developed a mathematical underpinning to a computer program that generated very complex and interesting art. But because he didn't know much math, his equations were bizarre—really ridiculous. He wanted the math to be beautiful but didn't know how to do it.

I was able to come up with a solution, not so complicated in physics at all: traveling waves in three dimensions. You can set the wavelength; you can determine the speed of the waves; and you can indicate their directions. And if you want three waves going through one another, you can do that. You start with a beginning condition and then you let the waves go through one another and add them up. This produces very interesting interference patterns.

The underlying math was beautiful, and that was very important for Peter. I don't mean to boast—he would tell you the same thing. This is the role that I have mostly played in his life: to show him how to make things mathematically beautiful and easy to understand. He very kindly always let me choose one work of art from each series. Lucky me, I have about thirteen Struyckens!

As a result of my collaboration with Peter, I was invited by the director of the Boijmans van Beuningen Museum in Rotterdam to give the first Mondrian Lecture in 1979 under the vast dome of Amsterdam's Koepelkerk. It was packed; there were about nine hundred people in my audience. This very prestigious lecture is now given every other year. The lecturer in 1981 was Umberto Eco, Donald Judd in 1993, Rem Koolhaas in 1995, and Charles Jencks in 2010.

My collaborations with Otto and Peter have not been my only involvement in making art; I once tried (in jest) to make a bit of conceptual art myself. When I gave my lecture "Looking at 20th-Century Art Through the Eyes of a Physicist" (http://mitworld.mit.edu/speaker/view/55), I explained that at home I have about a dozen books on physics but at least two hundred fifty on art, so the ratio is about twenty to one. I placed ten art books on the desk and invited the audience to look through them at the intermission. In order to keep the proper balance, I announced, I'd brought half a book on physics. That morning I had sliced a physics text in two, right down the middle of the spine. So I held it up, pointing out that I'd cut it very carefully—it was really half a book. "For those of you uninterested in art," I said—dropping it loudly on the table—"here you are!" I'm afraid no one got it.

If we look back at the days of Renaissance art up to the present, then there is a clear trend. The artists are gradually removing the constraints that were put on them by prevailing traditions: constraints of subject matter, of form, of materials, of perspective, of technique, and of color. By the end of the nineteenth century, artists completely abandoned the idea of art as a representation of the natural world.

The truth is that we now find many of these pioneering works magnificent, but the intention of the artists was quite something else. They wanted to introduce a new way of looking at the world. Many of the works that we admire today as iconic and beautiful creations—van Gogh's *Starry Night*, for example, or Matisse's *The Green Stripe* (a portrait of his wife) received ridicule and hostility at the time. Today's beloved Impressionists—Monet, Degas, Pissarro, Renoir—among the most popular artists in any museum today, also faced derision when they began showing their paintings.

The fact that most of us find their works beautiful now shows that the artists triumphed over their age: their new way of seeing, their new way of looking at the world, has become our world, our way of seeing. What was just plain ugly a hundred years ago can now be beautiful. I love the fact that a contemporary critic called Matisse the apostle of ugliness. The collector Leo Stein referred to his painting of Madame Matisse, *Woman with a Hat,* as "the nastiest smear I have ever seen"—but he bought the painting!

In the twentieth century artists used found objects—sometimes shocking ones, like Marcel Duchamp's urinal (which he called "fountain") and his Mona Lisa, on which he wrote the provocative letters L.H.O.O.Q. Duchamp was the great liberator; after Duchamp anything goes! He wanted to shake up the way we look at art.

No one can look at color in the same way after van Gogh, Gauguin, Matisse, and Derain. Nor can anyone look at a Campbell's soup can or an image of Marilyn Monroe in the same way after Andy Warhol.

Pioneering works of art can be beautiful, even stunning, but most often—certainly at first—they are baffling, and may even be ugly. The real beauty of a pioneering work of art, no matter how ugly, is in its meaning. A new way of looking at the world is never the familiar warm bed; it's always a chilling cold shower. I find that shower invigorating, bracing, liberating.

I think about pioneering work in physics in this same way. Once physics has taken another of its wonderfully revelatory steps into previously invisible or murky terrain, we can never see the world quite the same way again.

The many stunning discoveries I've introduced through this book were deeply perplexing at the time they were made. If we have to learn the mathematics behind those discoveries, it can be truly daunting. But I hope that my introduction of some of the biggest breakthroughs has brought to life just how exciting and beautiful they are. Just as Cézanne, Monet, van Gogh, Picasso, Matisse, Mondrian, Malevich, Kandinsky, Brancusi, Duchamp, Pollock, and Warhol forged new trails that challenged the art world, Newton and all those who have followed him gave us new vision.

The pioneers in physics of the early twentieth century—among them Antoine Henri Becquerel, Marie Curie, Niels Bohr, Max Planck, Albert Einstein, Louis de Broglie, Erwin Schrödinger, Wolfgang Pauli, Werner

Heisenberg, Paul Dirac, Enrico Fermi—proposed ideas that completely undermined the way scientists had thought about reality for centuries, if not millennia. Before quantum mechanics we believed that a particle is a particle, obeying Newton's laws, and that a wave is a wave obeying different physics. We now know that all particles can behave like waves and all waves can behave like particles. Thus the eighteenth-century issue, whether light is a particle or a wave (which seemed to be settled in 1801 by Thomas Young in favor of a wave—see chapter 5), is nowadays a non-issue as it is both.

Before quantum mechanics it was believed that physics was deterministic in the sense that if you do the same experiment a hundred times, you will get the exact same outcome a hundred times. We now know that that is not true. Quantum mechanics deals with probabilities—not certainties. This was so shocking that even Einstein never accepted it. "God does not throw dice" were his famous words. Well, Einstein was wrong!

Before quantum mechanics we believed that the position of a particle and its momentum (which is the product of its mass and its velocity) could, in principle, simultaneously be determined to any degree of accuracy. That's what Newton's laws taught us. We now know that that is not the case. Nonintuitive as this may be, the more accurately you can determine its position, the less accurately can you determine its momentum; this is known as Heisenberg's uncertainty principle.

Einstein argued in his theory of special relativity that space and time constituted one four-dimensional reality, spacetime. He postulated that the speed of light was constant (300,000 kilometers per second). Even if a person were approaching you on a superfast train going at 50 percent of the speed of light (150,000 kilometers per second), shining a headlight in your face, you and he would come up with the same figure for the speed of light. This is very nonintuitive, as you would think that since the train is approaching you, you who are observing the light aimed at you would have to add 300,000 and 150,000, which would lead to 450,000 kilometers per second. But that is not the case—according to Einstein, 300,000 plus 150,000 is still 300,000! His theory of general relativity was perhaps even more mind-boggling, offering a complete reinterpretation of the force holding the astronomical universe together, arguing that gravity functioned by distorting the fabric of spacetime itself, pushing bodies into orbit through geometry, even forcing light to bend through the same distorted spacetime. Einstein showed that Newtonian physics needed important revisions, and he opened the way to modern cosmology: the big bang, the expanding universe, and black holes.

When I began lecturing at MIT in the 1970s, it was part of my personality to put more emphasis on the beauty and the excitement rather than the details that would be lost on the students anyway. In every subject I taught I always tried where possible to relate the material to the students' own world—and make them see things they'd never thought of but were within reach of touching. Whenever students ask a question, I always say, "that's an excellent question." The absolute last thing you want to do is make them feel they're stupid and you're smart.

There's a moment in my course on electricity and magnetism that's very precious to me. For most of the term we've been sneaking up, one by one, on Maxwell's equations, the stunningly elegant descriptions of how electricity and magnetism are related—different aspects of the same phenomenon, electromagnetism. There's an intrinsic beauty in the way these equations talk to one another that is unbelievable. You can't separate them; together they're one unified field theory.

So I project these four beautiful equations on different screens on all the walls of the lecture hall. "Look at them," I say. "Inhale them. Let them penetrate your brains. Only once in your life will you see all four of Maxwell's equations for the first time in a way that you can appreciate them, complete and beautiful and talking to each other. This will never happen again. You will never be the same. You have lost your virginity." In honor of this momentous day in the lives of the students, as a way of celebrating the intellectual summit they've reached, I bring in six hundred daffodils, one for each student.

Students write me many years afterward, long after they've forgotten the details of Maxwell's equations, that they remember the day of the daffodils, the day I marked their new way of seeing with flowers. To me this is teaching at the highest level. It's so much more important to me for students to remember the beauty of what they have seen than whether they can reproduce what you've written on the blackboard. What counts is not what you cover, but what you uncover!

My goal is to make them love physics and to make them look at the world in a different way, and that is for life! You broaden their horizon, which allows them to ask questions they have never asked before. The point is to unlock the world of physics in such a way that it connects to the genuine interest students have in the world. That's why I always try to show my students the forests, rather than take them up and down every single tree. That is also what I have tried to do in this book for you. I hope you have enjoyed the journey.

ACKNOWLEDGMENTS

Without the intelligence, foresight, business sense, and moral support of our exceptional literary agent, Wendy Strothman, *For the Love of Physics* would have remained little more than wishful thinking. She brought the two of us together, found the right home for this book at Free Press, read numerous draft chapters with an editorial eye honed by her years as a publisher, gave the book its title, and helped keep us focused on the end product. We are also the happy and fortunate recipients of her staunch friendship, which buoyed us throughout the project.

It would be hard to overstate the contributions of our editor, Emily Loose, at Free Press, whose vision for this book proved infectious and whose extraordinarily close attention to prose narrative provided an education for both of us. Despite the enormous pressure in the publishing industry to cut corners on behalf of the bottom line, Emily insisted on really editing this book, pushing us always to greater clarity, smoother transitions, and tighter focus. Her skill and intensity have made this a far better book. We are grateful as well to Amy Ryan for her deft copyediting of the manuscript.

Walter Lewin:

Every day I receive wonderful, often very moving email from dozens of people all over the world who watch my lectures on the web. These lectures were made possible due to the vision of Richard (Dick) Larson. In

1998 when he was the director of the Center for Advanced Educational Services and a professor in the Department of Electrical Engineering at MIT, he proposed that my rather unconventional lectures be videotaped and made accessible to students outside MIT. He received substantial funding for this from the Lord Foundation of Massachusetts and from Atlantic Philanthropies. Dick's initiative was the precursor of e-learning! When MIT's OpenCourseWare opened its doors in 2001, my lectures reached all corners of the world and are now viewed by more than a million people each year.

During the past two years, even during the seventy days that I was in the hospital (and almost died), this book was always on my mind. At home I talked about it incessantly with my wife, Susan Kaufman. It kept me awake many nights. Susan patiently endured all this and managed to keep my spirits up. She also trained her astute editorial eye on a number of chapters and improved them markedly.

I am very grateful to my cousin Emmie Arbel-Kallus and my sister, Bea Bloksma-Lewin, for sharing with me some of their very painful recollections of events during World War II. I realize how difficult this must have been for both of them, as it was for me. I thank Nancy Stieber, my close friend for thirty years, both for always correcting my English and for her invaluable comments and suggestions. I also want to thank my friend and colleague George Clark, without whom I would never have become a professor at MIT. George let me read the original American Science and Engineering proposal submitted to the Air Force Cambridge Research Laboratories that led to the birth of X-ray astronomy.

I am grateful to Scott Hughes, Enectali Figueroa-Feliciano, Nathan Smith, Alex Filippenko, Owen Gingerich, Andrew Hamilton, Mark Whittle, Bob Jaffe, Ed van den Heuvel, Paul Murdin, Jeff McClintock, John Belcher, Max Tegmark, Richard Lieu, Fred Rasio, the late John Huchra, Jeff Hoffman, Watti Taylor, Vicky Kaspi, Fred Baganoff, Ron Remillard, Dan Kleppner, Bob Kirshner, Paul Gorenstein, Amir Rizk, Chris Davlantes, Christine Sherratt, Markos Hankin, Bil Sanford, and Andrew Neely for helping me, when help was needed.

Finally I can't thank Warren Goldstein enough for his patience with me and for his flexibility; at times he must have felt overwhelmed (and perhaps frustrated) with too much physics in too little time.

Warren Goldstein:

I would like to thank the following people for their willingness to talk with me about Walter Lewin: Laura Bloksma, Bea Bloksma-Lewin, Pauline Broberg-Lewin, Susan Kaufman, Ellen Kramer, Wies de Heer, Emanuel (Chuck) Lewin, David Pooley, Nancy Stieber, Peter Struycken. Even if they are not quoted in *For the Love of Physics,* each one added substantially to my understanding of Walter Lewin. Edward Gray, Jacob Harney, Laurence Marschall, James McDonald, and Bob Celmer did their best to keep Walter and me from making mistakes in their fields of expertise; as much as we'd prefer to put the onus on them, we take full responsibility for any remaining errors. I also want to thank William J. Leo, a 2011 graduate of the University of Hartford, for his assistance at a critical moment. Three of the smartest writers I know—Marc Gunther, George Kannar, and Lennard Davis—all gave me invaluable advice early in the project. In different ways Dean Joseph Voelker and Assistant Provost Fred Sweitzer of the University of Hartford made it possible for me to find the time to finish this book. I am deeply grateful to my wife, Donna Schaper—minister and organizer extraordinaire, and author of thirty books at last count—for understanding and celebrating my immersion in a foreign world. Our grandson, Caleb Benjamin Luria, came into the world October 18, 2010; it has been a delight to watch him undertake his own series of remarkable experiments in the physics of everyday life. Finally, I want here to express my deep gratitude to Walter Lewin, who taught me more physics in the last few years than either of us would have thought possible and rekindled a passion in me that had lain dormant far too long.

Mammal Femurs

I t's reasonable to assume that the mass of a mammal is proportional to its volume. Let's take a puppy and compare it with a full-grown dog that is four times bigger. I am assuming that all linear dimensions of the bigger dog are four times larger than that of the puppy—its height, its length, the length and the thickness of its legs, the width of its head, everything. If that is the case, then the volume (and thus the mass) of the bigger dog is about sixty-four times that of the puppy.

One way to see this is by taking a cube with sides a, b, and c. The volume of this cube is $a \times b \times c$. When you make all sides four times larger, the volume becomes $4a \times 4b \times 4c$, which is $64abc$. If we express this a bit more mathematically, we can say that the volume (thus the mass) of the mammal is proportional to its length to the third power. If the bigger dog is four times larger than the puppy, then its volume should be about 4 cubed (4^3) times larger, which is 64. So, if we call the length of the femur "l," then by comparing mammals of different size, their mass should be roughly proportional to l cubed (l^3).

Okay, that's mass. Now, the strength of the mammal's femur supporting all that weight has to be proportional to its thickness, right? Thicker bones can support more weight—that's intuitive. If we translate that idea to mathematics, the strength of the femur should be proportional to the area of the cross section of the bone. That cross section is roughly a circle, and we know that the area of a circle is πr^2, where r is the radius of the circle. Thus, the area is proportional to d^2 if d is the diameter of the circle.

Let's call the thickness of the femur "d" (for diameter). Then, following Galileo's idea, the mass of the mammal would be proportional to d^2 (so that the bones can carry the weight of the mammal), but it is also proportional to l^3 (that is always the case, independent of Galileo's idea). Thus, if Galileo's idea is correct, d^2 should be proportional to l^3, which is the same as stating that d is proportional to $l^{3/2}$.

If I compare two mammals and one is five times bigger than the other (thus the length l of its femur is about five times larger than that of the smaller mammal), I may expect that the thickness, d, of its femur is about $5^{3/2} = 11$ times greater than the thickness of the smaller animal's femur. In lectures I showed that the length l of the femur of an elephant was about 100 times larger than the length of the femur of a mouse; we may therefore expect, if Galileo's idea is correct, that the thickness, d, of the elephant's femur is about $100^{3/2} = 1,000$ times thicker than that of the mouse.

Thus at some point, for very heavy mammals, the thickness of the bones would have to be the same as their lengths—or even greater— which would make for some pretty impractical mammals, and that would then be the reason why there is a maximum limit on the size of mammals.

Newton's Laws at Work

Newton's law of universal gravitation can be written as

$$F_{grav} = G\frac{m_1 m_2}{r^2} \qquad [1]$$

Here, F_{grav} is the force of gravitational attraction between an object of mass m_1 and one of mass m_2, and r is the distance between them. G is called the gravitational constant.

Newton's laws made it possible to calculate, at least in principle, the mass of the Sun and some planets.

Let's see how this works. I'll start with the Sun. Suppose m_1 is the mass of the Sun, and that m_2 is the mass of a planet (any planet). I will assume that the planetary orbit is a circle of radius r and let the orbital period of the planet be T (T is 365.25 days for the Earth, 88 days for Mercury, and almost twelve years for Jupiter).

If the orbit is circular or nearly so (which is the case for five of the six planets known in the seventeenth century), the speed of a planet in orbit is constant, but the direction of its velocity is always changing. However, whenever the direction of the velocity of any object changes, even if there is no change in speed, there must be an acceleration, and thus, according to Newton's second law, there must be a force to provide that acceleration.

It's called the centripetal force (F_c), and it is always exactly in the direction from the moving planet toward the Sun. Of course, since Newton was Newton, he knew exactly how to calculate this force (I derive the equation in my lectures). The magnitude of this force is

$$F_c = \frac{m_2 v^2}{r} \qquad [2]$$

Here v is the speed of the planet in orbit. But this speed is the circumference of the orbit, $2\pi r$, divided by the time, T, it takes to make one revolution around the Sun. Thus we can also write:

$$F_c = \frac{4\pi^2 m_2 r}{T^2} \qquad [3]$$

Where does this force come from? What on earth (no pun implied) is the origin of this force? Newton realized that it must be the gravitational attraction by the Sun. Thus the two forces in the above equations are one and the same force; they are equal to each other:

$$F_{grav} = F_c \qquad [4]$$

If we massage this a bit further by rearranging the variables (this is your chance to brush up on your high school algebra), we find that the mass of the Sun is

$$m_1 = \frac{4\pi^2 r^3}{GT^2} \qquad [5]$$

Notice that the mass of the planet (m_2) is no longer present in equation 5; it does not enter into the picture; all we need is the planet's mean distance to the Sun and its orbital period (T). Doesn't that surprise you? After all, m_2 shows up in equation 1 and also in equation 2. But the fact that it is present in both equations is the very reason that m_2 is eliminated by setting F_{grav} equal to F_c. That's the beauty of this method, and we owe all this to Sir Isaac!

Equation 5 indicates that $\frac{r^3}{T^2}$ is the same for all planets. Even though they all have very different distances to the Sun and very different orbital periods, $\frac{r^3}{T^2}$ is the same for all. The German astronomer and mathematician Johannes Kepler had already discovered this amazing result in 1619, long before Newton. But why this ratio—between the cube of the radius and square of the orbital period—was constant was not understood at all. It was the genius Newton who showed sixty-eight years later that it is the natural consequence of his laws.

In summary, equation 5 tells us that if we know the distance from any planet to the Sun (r), the orbital period of the planet (T), and G, we can calculate the mass of the Sun (m_1).

Orbital periods were known to a high degree of accuracy long before the seventeenth century. The distances between the Sun and the planets were also known to a high degree of accuracy long before the seventeenth century but only on a *relative* scale. In other words, astronomers knew that Venus's mean distance to the Sun was 72.4 percent of Earth's and that Jupiter's mean distance was 5.200 times larger than Earth's. However, the absolute values of these distances were an entirely different story. In the sixteenth century, in the day of the great Danish astronomer Tycho Brahe, astronomers believed that the distance from the Earth to the Sun was twenty times smaller than what it actually is (close to 150 million kilometers, about 93 million miles). In the early seventeenth century Kepler came up with a more accurate distance to the Sun, but still seven times smaller than what it is.

Since equation 5 indicates that the mass of the Sun is proportional to the distance (to a planet) cubed, if the distance r is too low by a factor of seven, then the mass of the Sun will be too low by a factor of 7^3, which is 343—not very useful at all.

A breakthrough came in 1672 when the Italian scientist Giovanni Cassini measured the distance from the Earth to the Sun to an accuracy of about 7 percent (impressive for those days), which meant that the uncertainty in r^3 was only about 22 percent. The uncertainty in G was probably at least 30 percent. So my guess is that by the end of the seventeenth century the mass of the Sun may have been known to an accuracy no better than 50 percent.

Since the relative distances from the Sun to the planets were known to a high degree of accuracy, knowing the absolute distance from the Sun to the Earth to 7 percent accuracy meant that the absolute distances to the Sun of the other five known planets could also be calculated to that same 7 percent accuracy by the end of the seventeenth century.

The above method to calculate the mass of the Sun can also be used to measure the mass of Jupiter, Saturn, and the Earth. All three planets had known moons in orbit; in 1610 Galileo Galilei discovered four moons of Jupiter, now known as the Galilean moons. If m_1 is the mass of Jupiter, and m_2 the mass of one of its moons, then we can calculate the mass of Jupiter, using equation 5, in the same way that we can calculate the mass of the Sun, except that now r is the distance between Jupiter and its moon, and T is the orbital period of that moon around Jupiter. The four Galilean moons (Jupiter has sixty-three moons!) have orbital periods of 1.77 days, 3.55 days, 7.15 days, and 16.69 days.

Accuracies in distances and in G have greatly improved over time. By the nineteenth century G was known to about 1 percent accuracy. It is now known to an accuracy of about 0.01 percent.

Let me show you a numerical example. Using equation 5, let's calculate together the mass of the Earth (m_1) by using the orbit of our Moon (with mass m_2). To use equation 5 properly, the distance, r, should be in meters, and T should be in seconds. If we then use 6.673×10^{-11} for G, we get the mass in kilograms.

The mean distance to the Moon (r) is 3.8440×10^8 meters (about 239,000 miles); its orbital period (T) is 2.3606×10^6 seconds (27.32 days). If we plug these numbers into equation 5, we find that the mass of the Earth is 6.030×10^{24} kilograms. The best current value of Earth's mass is close to 5.974×10^{24} kilograms, which is only 1 percent lower than what I calculated! Why the difference? One reason is that the equation we used assumed that the Moon's orbit is circular, when in fact it is elongated, what we call elliptical. As a result, the smallest distance to the Moon is about 224,000 miles; the largest is about 252,000 miles. Of course, Newton's laws can also easily deal with elliptical orbits, but the math may blow your mind. Perhaps it already has!

There is another reason why our result for the mass of the Earth is a little off. We assumed that the Moon circles around the Earth and that the center of that circle is the center of the Earth. Thus in equations 1 and 3, we assumed that r is the distance between the Earth and the Moon. That is correct in equation 1; however, as I discuss in more detail in chapter 13, the Moon and the Earth actually each orbit the center of mass of the Moon-Earth system, and that is about a thousand miles below the Earth's surface. Thus r, in equation 3, is a little less than r in equation 1.

Since we live on Earth, there are other ways of calculating the mass of our home planet. One is by measuring the gravitational acceleration near the surface. When dropped, any object of mass m (m can have any value) will be accelerated with an acceleration, g, close to 9.82 meters per second per second.* Earth's average radius is close to 6.371×10^6 meters (about 3,960 miles).

Now let's revisit Newton's equation 1. Since $F = ma$ (Newton's second law), then

$$G\frac{m_{earth}m}{r^2} = mg \qquad [6]$$

Here, r is the radius of the Earth. With $G = 6.673 \times 10^{-11}$, $g = 9.82$ meters per second per second, and $r = 6.371 \times 10^6$ meters, we can calculate m_{earth} in kilograms (you try it!). If we simplify equation 6 somewhat, we get

$$m_{earth} = \frac{gr^2}{G} \qquad [7]$$

I find that m_{earth} is 5.973×10^{24} kilograms (impressive, right?).

Notice that the mass, m, of the object we dropped does not show up

---

*This acceleration, by the way, is 0.18 percent lower at the equator than at the poles—because Earth is not a perfect sphere. Objects at the equator are about 20 kilometers farther away from the Earth's center than objects at the poles, so at the equator g is lower. The 9.82 is an average value.

in equation 7! That should not surprise you, as the mass of the Earth could not possibly depend on the mass of the object that you drop.

You might also be interested in knowing that Newton believed that the average density of the Earth was between 5,000 and 6,000 kilograms per cubic meter. This was not based on any astronomical information; it was completely independent of any of his laws. It was his best "educated" guess. The average density of the Earth is, in fact, 5,540 kilograms per cubic meter. If you allow me to write Newton's guess as 5,500 ± 500 kilograms per cubic meter, his uncertainty was only 10 percent (amazing!).

I do not know if Newton's guess was ever taken seriously in his day, but suppose it was. Since Earth's radius was well known in the seventeenth century, its mass could have been calculated to an accuracy of 10 percent (mass is volume times density). Equation 7 could then be used to calculate G also to an accuracy of 10 percent. I am telling you this because it intrigues me that, accepting Newton's guess for the mean density of the Earth, at the end of the seventeenth century the gravitational constant, G, could already have been known to an accuracy of 10 percent!

INDEX

ABOUT THE AUTHORS

Walter Lewin was born and raised in the Netherlands. In 1965 he received his PhD in Physics from the University of Technology in Delft. He arrived at MIT in 1966 as a postdoctoral fellow. That same year he became an assistant professor and in 1974 was made full professor. He is a highly accomplished astrophysicist, a pioneer in X-ray astronomy, and has published more than four hundred and fifty scientific articles. Lewin taught the three physics core classes at MIT for more than thirty years. These lectures were so popular that they were videotaped and became hits on MIT's OpenCourseWare, YouTube, iTunes U, and Academic Earth. More than a million people from all over the world watch these lectures yearly. Acclaim for his lectures has been featured in many media outlets, including the *New York Times, Boston Globe, International Herald Tribune, Guardian, Washington Post, Newsweek,* and *U.S. News and World Report.* His honors and awards include the NASA Exceptional Scientific Achievement Medal (1978), the Alexander von Humboldt Award, a Guggenheim Fellowship (1984), MIT's Science Council Prize for Excellence in Undergraduate Teaching (1984), the W. Buechner Prize of the MIT Department of Physics (1988), the NASA Group Achievement Award for the Discovery of the Bursting Pulsar (1997), and the Everett Moore Baker Memorial Award for Excellence in Undergraduate Teaching (2003). He became a corresponding member of the Royal Netherlands Academy of Arts and Sciences and Fellow of the American Physical Society in 1993.

Warren Goldstein is a professor of history and chair of the History Department at the University of Hartford, where he was the recipient of the James E. and Frances W. Bent Award for Scholarly Creativity (2006). He has a lifelong fascination with physics. A prolific and prizewinning historian, essayist, journalist, and lecturer, his prior books include *Playing for Keeps: A History of Early Baseball,* and the critically acclaimed biography *William Sloane Coffin, Jr.: A Holy Impatience.* His writing about history, education, religion, politics, and sports has appeared in the *New York Times, Washington Post, Chronicle of Higher Education, Boston Globe, Newsday, Chicago Tribune, Philadelphia Inquirer, Nation, Christian Century, Yale Alumni Magazine, Times Literary Supplement,* and *Huffington Post.*